北部湾大学项目教学系列

ArcGIS Runtime for.NET 开发实验实习教程
——基于C#和WPF

主　编　田义超　谢小魁　魏金占　林　卉
副主编　黄远林　吴　宁　丁　峰　杨芸珍
　　　　邢鹏威　孟子源　马咸福　张　强
　　　　陶　进

武汉大学出版社

图书在版编目(CIP)数据

ArcGIS Runtime for .NET 开发实验实习教程:基于 C#和 WPF/田义超等主编. —武汉:武汉大学出版社,2022.12
北部湾大学项目教学系列
ISBN 978-7-307-23180-1

Ⅰ.A… Ⅱ.田… Ⅲ.地理信息系统—应用软件—软件开发—教材 Ⅳ.P208

中国版本图书馆 CIP 数据核字(2022)第 133096 号

责任编辑:谢文涛　　责任校对:汪欣怡　　版式设计:马　佳

出版发行:武汉大学出版社　　(430072　武昌　珞珈山)
(电子邮箱:cbs22@whu.edu.cn 网址:www.wdp.com.cn)
印刷:武汉中科兴业印务有限公司
开本:787×1092　1/16　印张:16　字数:376 千字　插页:1
版次:2022 年 12 月第 1 版　　2022 年 12 月第 1 次印刷
ISBN 978-7-307-23180-1　　定价:46.00 元

版权所有,不得翻印;凡购买我社的图书,如有质量问题,请与当地图书销售部门联系调换。

前　　言

随着地理信息科学和 IT 技术的迅速发展，传统的地理信息开发方式 MapObjects、ArcObjects、ArcGIS Engine 存在部署困难、不能跨平台、开发难度大、成本高、效率低、性能差等缺点，急需向互联网+大数据+云计算+跨平台的现代开发方式转变和升级。

ArcGIS Runtime 是用现代开发理念打造的跨移动和桌面平台的全新开发组件，具有不用安装、复制即用、无 COM、无 Python、纯净优雅、功能强大等优点。

本书基于现代 GIS 体系结构和理念，精心设计并开发了数十个应用程序（Apps）实例，通过案例系统地介绍如何使用 Visual Studio 和 ArcGIS Runtime SDk 开发 .NET WPF（Windows Presentation Foundation）独立 Apps，涉及主流数据类型和数据格式（shp、lpk、tpk、mmpk、GeoPackage、geodatabase、online、server、portal）、在线和离线、二维和三维、显示和分析等。主要内容包括高质量高性能的交互制图、空间和属性查询、地理编码、数据编辑和高级地理分析。本书适合作为地理信息、测绘、遥感、软件工程、物联网等相关专业高年级本科生和研究生的 GIS 开发教材，也可以作为软件开发工程技术人员的参考用书。

学习本书要求具有初步的 C#编程、WPF 开发和 GIS 操作基础，不需要具有 ArcGIS 开发基础。

目 录

第1章 ArcGIS Runtime 入门和地图显示 1
- 第1节 ArcGIS Runtime 入门 1
- 第2节 ArcGIS Runtime SDK 安装 3
- 第3节 零代码创建第一个地图应用程序 5
- 第4节 异步编程入门 9
- 第5节 底图浏览器 13
- 第6节 在线网络底图选择器 14
- 第7节 URL 在线网络地图浏览器 17
- 第8节 门户地图浏览器 20

第2章 常见图层类型 23
- 第1节 ArcGIS 地图图像图层浏览器 23
- 第2节 ArcGIS 地图图像图层可见性 25
- 第3节 在线 ArcGIS 切片图层浏览器 27
- 第4节 ArcGIS 矢量切片图层浏览器 28
- 第5节 要素集合图层门户浏览器 31
- 第6节 网络地图服务浏览器 33

第3章 离线地图 36
- 第1节 移动地图包浏览器 36
- 第2节 矢量图层地理包浏览器 38
- 第3节 栅格图层地理包浏览器 40
- 第4节 TPK 浏览器 43
- 第5节 矢量切片包浏览器 45
- 第6节 SHP 浏览器 46
- 第7节 栅格文件浏览器 48
- 第8节 栅格地理包浏览器 50
- 第9节 影像服务栅格浏览器 53
- 第10节 多光谱影像彩色渲染器 55
- 第11节 栅格拉伸渲染器 59

第 4 章　在线三维场景浏览 ·· 68
- 第 1 节　高程服务场景三维浏览器 ·· 68
- 第 2 节　在线场景三维浏览器 ·· 70
- 第 3 节　门户场景三维浏览器 ·· 73
- 第 4 节　Url 场景浏览器 ··· 75
- 第 5 节　查看要素图层属性表 ·· 77
- 第 6 节　查看要素属性 ·· 80
- 第 7 节　要素标注控制器 ·· 83

第 5 章　查询和编辑 ·· 87
- 第 1 节　要素图层查询器 ·· 87
- 第 2 节　Shp 文件查询器 ·· 90
- 第 3 节　Sql 查询器 ··· 93
- 第 4 节　识别图层要素 ·· 95
- 第 5 节　识别符号 ·· 98
- 第 6 节　绘制点 ··· 101
- 第 7 节　绘制线 ··· 103
- 第 8 节　绘制多边形 ··· 106
- 第 9 节　点击选择要素 ··· 108
- 第 10 节　画线选择要素 ·· 110
- 第 11 节　画面选择要素 ·· 113
- 第 12 节　属性编辑 ·· 116
- 第 13 节　Shp 点要素编辑 ·· 119
- 第 14 节　SHP 线要素编辑 ·· 122
- 第 15 节　Shp 多边形要素编辑 ·· 125
- 第 16 节　要素集合图层查询器 ·· 128
- 第 17 节　要素集合图层构建器 ·· 130

第 6 章　实用工具篇 ··· 135
- 第 1 节　地图坐标查看器 ··· 135
- 第 2 节　图层视图状态查看器 ··· 137
- 第 3 节　坐标投影转换器 ··· 138
- 第 4 节　shp 工作空间读取器 ··· 140
- 第 5 节　地名标注器 ··· 142
- 第 6 节　设备地址显示 ··· 144
- 第 7 节　设备地址坐标读取器 ··· 147
- 第 8 节　进度查看器 ··· 149
- 第 9 节　服务要素表手动缓存管理器 ······································· 151

第 10 节　移动地理数据库浏览器 ··· 153
  第 11 节　移动地理数据库下载器 ··· 155
  第 12 节　时间范围浏览器 ·· 161
  第 13 节　路径规划 ·· 164
  第 14 节　图层范围合并器 ·· 169

第 7 章　显示控制与渲染 ··· 173
  第 1 节　网格控制器 ··· 173
  第 2 节　要素图层生成符号 ··· 175
  第 3 节　栅格山体阴影渲染器 ·· 178
  第 4 节　电子海图浏览器 ·· 182
  第 5 节　军事符号地图 ··· 186
  第 6 节　视点管理器 ··· 190
  第 7 节　模拟飞行三维动画 ··· 192

第 8 章　地理信息处理服务 ··· 199
  第 1 节　基于位置的三维视线分析 ··· 199
  第 2 节　基于图形的三维视线分析 ··· 203
  第 3 节　基于位置的三维视域分析 ··· 208
  第 4 节　基于服务的最近设施分配 ··· 215
  第 5 节　基于本地数据的最近设施分配 ·· 219
  第 6 节　本地数据的服务范围分配 ··· 224
  第 7 节　等高线生成器 ··· 230

第 9 章　项目教学与实习案例 ·· 237
  项目 1　畜禽养殖污染管理地理信息系统 ·· 237
  项目 2　森林经营决策地理信息系统 ·· 238
  项目 3　三湘四水——湖南地貌水系三维地图秀 ··· 239
  项目 4　海产品溯源监测地理信息系统 ··· 239
  项目 5　POI 标注地理信息系统 ··· 240
  项目 6　停车场智能管理地理信息系统 ··· 241
  项目 7　不动产登记地理信息系统 ··· 242
  项目 8　通用地理信息平台 ··· 243
  项目 9　通用三维地理信息系统 ·· 244
  项目 10　万能地图下载器 ··· 244
  项目 11　基于深度学习的智能解译系统 ··· 245
  项目 12　移动导航 ··· 245
  项目 13　遥感解译外业数据采集 ··· 245

项目 14　出租车监控服务端和客户端　　246
项目 15　导游导览导流预警智慧景区　　246
项目 16　公共设施管理系统　　247

参考文献　　248

第1章　ArcGIS Runtime 入门和地图显示

随着计算机信息技术和地理信息科学的快速发展，地理信息系统 GIS(Geographical Information System)已进入千家万户，成为百姓生活、社会生产和科学研究中不可或缺的基本工具。二次开发是 GIS 适配行业应用的主要方式，是其顽强生命力和应用多样性的重要体现。ArcGIS 是美国 ESRI 公司的主要产品，以技术可靠、算法先进、科研引领而著称于世，在软件体系结构和空间数据库方面一直引导着 GIS 行业的发展，其发展基本代表了国际上地理信息科学和技术的前沿水平。

ArcGIS 是一个全面的 GIS 平台，具有多种二次开发方式，为行业应用提供了丰富多样的定制手段。其传统的二次开发方式主要基于脚本和 COM 组件技术。传统的脚本语言包括 Avenue 和 AML 脚本语言；COM 组件包括小型 MapObjects、细粒度 ArcObjects(AOs)和粗粒度 ArcGIS Engine，曾经为 ArcGIS 的行业应用和普及提供了强大的技术支持。但 ArcGIS 专用的脚本语言存在功能弱、安全性差、应用范围窄等缺点；而 COM 组件由于设计理念落后、版本控制难(例如，ArcGIS 桌面软件各版本不能同时安装在一台电脑上，ArcObjects、ArcGIS Engine 和 ArcGIS Desktop 版本必须完全一致)，还存在注册表地狱、不能跨平台、非类型安全、开发难度大、部署困难等系列问题，急需向互联网+大数据+云计算+跨平台的现代开发方式转变和升级。

ArcGIS Runtime 二次开发采用了全新的设计理念，具有轻量级、绿色免安装(即插即用)、可视化、云计算、移动计算、跨平台等特点。对开发者而言，具有开发实例众多、辅助工具丰富、有成功案例可供借鉴等优点。在 ArcGIS Runtime 平台上开发的 ArcGIS Earth 就是很好的案例。ArcGIS Earth 是免费的交互式三维数字地球平台，可以使用 ESRI 公司的全球地形、影像服务，足不出户探索地球；还可以添加自定义的网络数据和本地数据，实现在三维空间内的数据展示和分析功能。

学完本章，读者将会在认识 ArcGIS Runtime 的基础上，熟悉在线、离线二维地图的显示。

第1节　ArcGIS Runtime 入门

本章介绍 ArcGIS Runtime for WPF SDK 的特点、安装，以及通过向导零代码创建 App 和异步编程的基本知识。

1. ArcGIS Runtime 功能

ArcGIS Runtime SDK for .NET 具有如下功能：

(1) 可以显示各种来源的丰富的交互地图；
(2) 添加本地设备上的要素、栅格和地图；
(3) 标准而强大的空间分析；
(4) 使用复合的空间、属性和时态标准，对地理特征进行查询、检索和识别；
(5) 网络计算、路径优化和导航；
(6) 在野外收集和编辑数据，与企业数据库同步；
(7) 跨平台，重用逻辑层代码。

2. ArcGIS Runtime SDK 的平台差异

ArcGIS Runtime SDK 包括平台特定的 API（WPF，UWP，Xamarin.Android，Xamarin.iOS，Xamarin Forms）。各平台版本特点和差异见表 1-1。

表 1-1　　　　　　　　**ArcGIS Runtime 支持的平台和设备**

ArcGIS Runtime 版本	支持平台和设备	开发语言
Android	Android	Java
iOS	Apple iPhone、iPod touch、iPad	Swift、Objective-C
Java	Windows、Linux	Java
.NET	Windows desktops、Windows Phone、Windows Store	.NET 兼容语言（如 C#、VB.NET）
OS X	Macs	Swift、Objective-C
Qt	Windows、Linux、Android、iOS	Qt
Xamarin	Android and iOS	C#

从表 1-1 可看出，ArcGIS Runtime 直接支持现代编程语言（如 C#、Java、Objective-C、Qt），原生支持主流操作系统（如 Windows 系列、Linux 系列）和移动平台（如 Android、iPhone 等）。这些系统的 API 都基于同一个 C++ 内核，使得各平台的接口风格和编程模型相一致；并在不同平台上做了对应的封装，又具有原生 API 的特点，因此开发流畅，避免了互操作所带来的性能损失和阻抗失配；这与超图的 SuperMap GIS Universal 共相式地理信息系统的理念一致。

ArcGIS Runtime 是 ESRI 公司基于现代软件理念开发的 GIS 二次开发包，摆脱了过去 COM 组件的局限，具有类型安全、跨平台、轻量级、云计算、异步调用、渲染快速、性能优越等现代 GIS 开发包的特点，是未来利用 ArcGIS 开发独立应用程序的主流模式。ArcGIS Runtime 的主要功能包括在离线一体化（online or offline）、二三维一体化、地图编辑、地理编码（geocoding）、路径导航和可视化。ArcGIS Runtime 具有开发简单、灵活多样、体验良好、容易部署等特点，适用于云计算和大数据背景下的行业应用。

3. ArcGIS Runtime 支持的数据类型

ArcGIS Runtime 支持的数据类型丰富，矢量和栅格数据类型都可以用作数据源。难能可贵的是，经典的 SHP 文件可以直接用于显示和分析，同时针对移动平台进行了定制和优化，可以提高时空性能。ArcGIS Runtime 支持的常见数据格式见表 1-2。

表 1-2　　　　　　　　**ArcGIS Runtime 支持的常见数据格式**

数据格式	英文名称	扩展名	支持矢量	支持栅格	备注
切片包	tile package	tpk	X	V	
矢量切片包	vector tile package	vtpk	V	X	
移动地理数据库	mobile geodatabase	geodatabase	V	/	
地理包	GeoPackage	gpkg	V	V	OGC 标准
图形文件	shapefile	shp	V	X	公开格式
栅格文件	raster file	tif, jpg…	X	V	
移动地图包	Mobile map package	mmpk	V	V	
图层包	Layer package	lpk	V	V	

第 2 节　ArcGIS Runtime SDK 安装

ArcGIS Runtime 开发包(Software Development Kit, SDK)在开发前需要安装。

可以通过 ESRI 开发账号免费下载安装包进行安装，下载前需要登录。如果没有账号，可以免费注册。

在同一台电脑上可以安装多个版本的应用程序开发包(SDK)。

开发 ArcGIS Runtime Apps 不需要许可，在任何设备和机器上学习和开发也不需要授权。用户可以免费加入 ArcGIS 开发者计划(Developer Program)，免费订阅，可获得所有功能的 APIs，用于开发目的。

WIN 平台的安装包为 Visual Studio 扩展(.vsix 文件)，包含项目模板和 NuGet 包。

1. 下载 SDK

ArcGIS Runtime SDK for .NET 下载地址为：

https://developers.arcgis.com/net/，下载界面见下图。

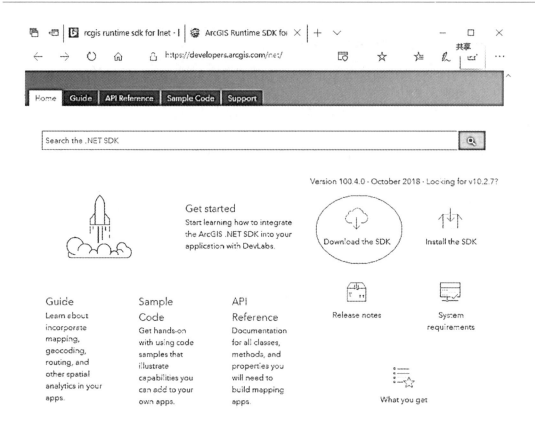

2. 安装 SDK

安装包会自动检测支持的 Visual Studio 环境，添加对应的扩展，安装界面如下：

3. 查看 SDK 安装结果

在 Visual Studio 中，点击菜单/Tools/Extensions and updates：

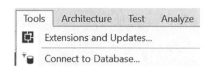

查看安装结果，如下图：

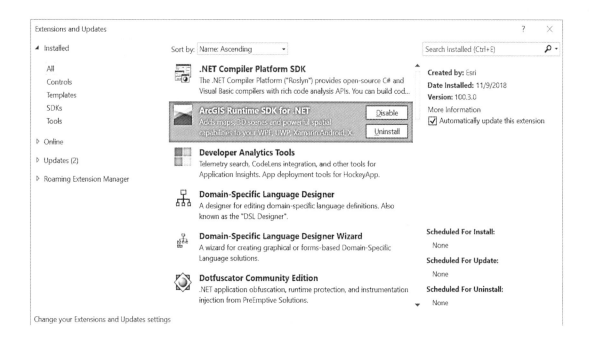

4. 卸载 SDK

在上一步的确认安装结果界面，选中扩展"ArcGIS Runtime SDK for .NET"后，点击"Uninstall"，即可卸载，操作界面如下图：

第 3 节　零代码创建第一个地图应用程序

对于大部分 ArcGIS Runtime SDK 应用程序，最核心的是 MapView 控件和对应的 Map。

【实验目的】

本实验以最为快速简单的方式，零代码创建第一个可以运行的地图应用程序 HelloArcGISApp，并查看解决方案和程序集信息。

【实验数据】

默认的全球街道网络底图 Basemap.CreateStreets()，需要网络连接。

【实验步骤】

1. 新建项目

.NET 框架：选择对应的版本，例如 ArcGIS Runtime 100.3 对应 .NET Framework 4.5.2，ArcGIS Runtime 100.4 对应 .NET Framework 4.6.1

模板列表：Visual C#/Windows/Windows Classic Desktop

模板：ArcGIS Runtime Application（WPF）

名称：HelloArcGISApp

位置：D:\ArcGISRuntimeTutorial

参数选择见下图：

新建 ArcGIS Runtime Application

2. 编译应用程序

MainWindow.xaml 自动显示出来，但有可能不能正确显示控件，还可能显示标记错误，见下图：

标记错误图

通过菜单 Build/Build Solution(快捷键 Ctrl+Shift+B)进行编译。编译(Build)应用程序后,将会自动修复标记错误,MapView 控件会正确显示全球地图轮廓,见下图:

编译结果图

3. 运行结果

用标准工具栏的 Start 按钮,或者菜单 Debug/Start Debugging,启动应用程序,正确显示网络底图，见下图：

运行结果图

4. 查看解决方案和程序集

在解决方案中，仔细查看包含的内容，确认 References 中引用的模块，重点关注 System 和 ESRI.ArcGISRuntime，查看对应的属性，见下图：

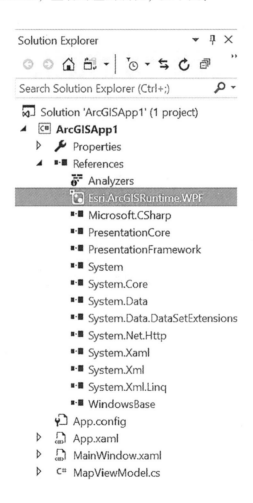

点击/Esri.ArcGISRuntime 上右键，选中属性，可查看程序集的信息（注意不同版本的程序集，显示信息略有差异）：

【实验讲解】

(1) 使用 ArcGIS Runtime Application 模板创建项目，将会自动添加合适的引用，在主窗口中将会加入 MapView 地图视图控件，并且自动加载底图。

(2) 新建的 WPF 程序中，包含一个 WPF 窗口，称为 MainWindow.xaml，其中定义了地图视图控件 MapView，用于显示地图 Map。

(3) 下面的 XAML 语句，通过数据绑定（Data binding）显示地图。

```
<Grid>
    < esri:MapView Map = "{Binding Map, Source = {StaticResource MapViewModel}}" />
</Grid>
```

(4) MapViewModel 包含了生成地图的逻辑，并且公开了属性 Map 给地图视图控件 MapView 使用。这是 MVVM（Model-View-ViewModel）设计模式的一个例子，用于分离业务逻辑和用户界面层。

(5) 默认情况下，地图显示第一个（最底层）图层的全部空间范围。如果需要显示特定的初始视图（initial view），需要在应用程序启动时，添加逻辑定义特定范围。

第 4 节　异步编程入门

I/O 密集型（如 Web 地图、文件、音频、视频、大数据）和计算密集型（如 GIS 空间分析、RS 遥感解译）应用，访问时可能会很慢或有延时，应用程序长时间没有响应，导致用户错误地认为程序已经失败。

通过使用异步编程，可以避免性能瓶颈，增强响应能力。C#中的 Async 和 Await 关键字是异步编程的核心。使用修饰符 async 关键字定义"异步方法"。按照约定，异步方法的名称以"Async"后缀结尾。方法通常包含至少一个 await 表达式，该表达式标记一个点，在该点上，直到等待的异步操作完成，方法才能继续。同时，将方法挂起，并且控件返回到方法的调用方。

异步返回类型为下列类型：

（1）如果方法有操作数为 TResult 类型的返回语句，则为 Task<TResult>；

（2）如果方法没有返回语句或具有没有操作数的返回语句，则为 Task；

（3）void：如果要编写异步事件处理程序。

【实验目的】

编写异步入门爬虫 HelloAsync，抓取 ESRI 主页 www.esri.com 的源代码 HelloAsync。

【实验数据】

www.esri.com，需要网络连接。

【实验步骤】

1. 新建 Windows Classic Desktop/WPF App 项目

项目名称为 HelloAsync，其他设置如下图：

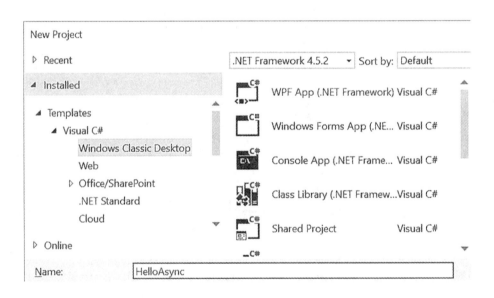

2. 修改 MainWindow.xaml 标记

设计程序运行后的 UI 如下：

为了实现上述 UI，把网格分为 3 行×1 列。

行号	控件	名称	目的
第一行	TextBox	tbUrl	输入网址
第二行	Button	StartButton	抓取网页内容(异步)
第三行	TextBox	resultsTextBox	显示网页内容

上述设计对应的 xaml 语言如下：

```xaml
<Grid>
    <Grid.RowDefinitions>
        <RowDefinition Height="Auto"/>
        <RowDefinition Height="Auto"/>
        <RowDefinition Height="*"/>
    </Grid.RowDefinitions>
    <TextBox Name="tbUrl" Grid.Row="0"/>
    <Button Name="StartButton" Content="Start" Click="StartButton_Click" Grid.Row="1"/>
    <TextBox Name="resultsTextBox" HorizontalScrollBarVisibility="Auto" VerticalScrollBarVisibility="Auto" Grid.Row="2"/>
</Grid>
```

3. 修改 MainWindow.xaml.cs 后台代码

添加命名空间：

```csharp
using System.Net.Http;
using System.Threading.Tasks;
using System.Windows;
```

编写 Initialize 函数，存储 ESRI 网址：

```csharp
void Initialize()
{
    tbUrl.Text = "http://www.esri.com";
}
```

在构造函数中调用 Initialize 函数进行初始化：

```
public MainWindow()
{
    InitializeComponent();
    Initialize ();
}
```

添加爬虫异步函数：

```
async Task<string> AccessTheWebAsync()
{
    HttpClient client = new HttpClient();
    string urlContents = await client.GetStringAsync(tbUrl.Text);
    return urlContents;
}
```

点击 Start 按钮，通过异步方式，读取网页源代码，并将源代码长度和内容显示在文本框中，代码为：

```
async void StartButton_Click(object sender, RoutedEventArgs e)
{
    resultsTextBox.Text = "Working .......\r\n";
    string content = await AccessTheWebAsync();
    resultsTextBox.Text +=
       $"Length of the downloaded string: {content.Length}.\r\n";
    resultsTextBox.Text += content;
}
```

【实验结果】

运行程序，将正确读取并显示 ESRI 网址的 HTML 源代码，如下图所示：

第 5 节　底图浏览器

ESRI 云计算提供了大量实用的在线网络底图，可以直接在 ArcGIS Runtime 中使用。

【实验目的】

编写底图浏览器程序 BasemapViewer，手动删除默认地图，手动加载底图，以增进对在线网络底图的认识，并分析不同底图之间的区别。

【地图数据】

默认的全球街道网络底图 Basemap，需要网络连接。

【实验步骤】

1. 新建项目

.NET 框架：与 ArcGIS Runtime SDK 对应
模板列表：Visual C#/Windows/Windows Classic Desktop
模板：ArcGIS Runtime Application（WPF）
名称：BasemapViewer
位置：D：\ArcGISRuntimeTutorial

2. 编译应用程序

详情见第 1 章第 3 节。

3. 修改地图视图

更改地图视图在网格中的行号，添加名称：
`<esri:MapView Grid.Row="1" Name="mapView1" />`

4. 修改 MainWindow.xaml.cs 代码

进入后台代码，在 MainWindow.xaml 右键，选择 View Code，进入 MainWindow.xaml.cs。

更改底图为带标注的影像：
```
public void Init()
{
    mapView1.Map.Basemap = Basemap.CreateImageryWithLabels();
}
```

在构造函数中调用初始化代码进行初始化
```
public MainWindow()
{
    InitializeComponent();
    Init();
```

}
【实验结果】

运行 App，地图将切换至对应的底图。

第 6 节　在线网络底图选择器

ESRI 提供了大量实用的在线网络底图，可以直接在 ArcGIS Runtime 中使用。

【实验目的】

本实验将浏览多个底图，以增进对在线网络底图的认识，并分析不同底图之间的区别。点击按钮 Button 切换不同的底图。

【地图数据】

ESRI 网络底图 Basemap，需要网络连接。

【实验步骤】

1. 新建项目

.NET 框架：与 ArcGIS Runtime SDK 对应

模板列表：Visual C#/Windows/Windows Classic Desktop

模板：ArcGIS Runtime Application（WPF）

名称：BasemapSelector

位置：D：\ArcGISRuntimeTutorial

2. 编译应用程序

见第 1 章第 3 节。

3. 修改 MainWindow.xaml 标记语言

网格设置。在标签 Grid 里，修改 XAML 代码，将网格设置为 2 行：

```
<Grid.RowDefinitions>
    <RowDefinition Height = "auto"/>
    <RowDefinition Height = " * "/>
</Grid.RowDefinitions>
```

插入工具条。在网格第一行插入工具条：

```
<ToolBar Grid.Row = "0">
</ToolBar>
```

加入命令按钮。在工具条加入命令按钮，用于确认用户的底图选择，随后修改后台代码，在点击事件中更改底图：

```
<Button Content = "CreateStreets" Name = "btCreateStreets" Click = "btCreateStreets_Click"/>
```

修改地图视图。更改地图视图在网格中的行号，添加名称：

```
<esri:MapView Grid.Row = "1" Name = "mapView1" Map = "{Binding Map, Source = {StaticResource MapViewModel}}"/>
```

4. 修改 MainWindow.xaml.cs 后台代码

在 MainWindow.xaml 右键，选择 View Code，进入 MainWindow.xaml.cs

通过点击事件，切换底图：

```
private void btCreateStreets_Click(object sender, RoutedEventArgs e)
{
    mapView1.Map.Basemap = Basemap.CreateStreets();
}
```

在工具条上再添加 3 个按钮，在点击事件中分别添加 OpenStreetMap，ImageryWithLabels，Topographic 等 3 种底图。

【实验结果】

点击上层按钮后，地图将切换至对应的底图：

【实验讲解】

(1) 开发者可以通过智能提示，自行查看 APIBasemap.Create***() 的所有方法，逐个添加按钮点击事件，进行底图切换，判断不同底图的差异，分析对应的应用场景。

(2) 本程序最终 MainWindow.xaml 主要代码：

```
<Grid.RowDefinitions>
    <RowDefinition Height="auto"/>
    <RowDefinition Height="*"/>
</Grid.RowDefinitions>
<ToolBar Grid.Row="0">
     <Button Content="CreateStreets" Name="btCreateStreets" Click="btCreateStreets_Click"/>
     <Button Content="CreateOpenStreetMap" Name="btCreateOpenStreetMap" Click="btCreateOpenStreetMap_Click"/>
     <Button Content="CreateImageryWithLabels" Name="btCreateImageryWithLabels" Click="btCreateImageryWithLabels_Click"/>
     <Button Content="CreateTopographic" Name="btCreateTopographic" Click="btCreateTopographic_Click"/>
</ToolBar>
<esri:MapView Grid.Row="1" Name="mapView1" Map="{Binding Map,Source={StaticResource MapViewModel}}"/>
```

(3) 本程序最终 MainWindow.xaml.cs 主要代码：

```
private void btCreateStreets_Click(object sender, RoutedEventArgs e)
{
    mapView1.Map.Basemap = Basemap.CreateStreets();
}
private void btCreateOpenStreetMap_Click(object sender, RoutedEventArgs e)
{
    mapView1.Map.Basemap = Basemap.CreateOpenStreetMap();
}
private void btCreateImageryWithLabels_Click(object sender, RoutedEventArgs e)
{
    mapView1.Map.Basemap = Basemap.CreateImageryWithLabels();
}
private void btCreateTopographic_Click(object sender, RoutedEventArgs e)
{
```

```
mapView1.Map.Basemap = Basemap.CreateTopographic();
}
```

第 7 节　URL 在线网络地图浏览器

地图(Map)指明了地理数据的组织方式以及如何与用户进行交互。地图实例必须分配到地图视图(MapView)上才能显示。在 MVC(Model View Controller)架构中,地图表示模型层(model tier),地图视图表示视图层(view tier)。地图和地图视图共同工作,将地理数据显示在屏幕上。

【实验目的】

本实验将通过按钮(Button)的点击事件,利用 URL 加载飓风和热带气旋底图。

【地图数据】

飓风和热带气旋网络底图(Basemap),需要网络连接。

地图视图链接,在 ArcGIS.com 地图浏览器中显示或创建地图:

https://www.arcgis.com/home/webmap/viewer.html? webmap = 69fdcd8e40734712aaec34194d4b988c

项目具体链接,可以在 ArcGIS.com 中查看详细页面:

https://www.arcgis.com/home/item.html? id=69fdcd8e40734712aaec34194d4b988c

项目数据链接,用 JSON 表达的地图显示信息:

https://www.arcgis.com/sharing/rest/content/items/69fdcd8e40734712aaec34194d4b988c/data?

通过 URL 在浏览器中打开,可以看到数据的介绍如下:

This map features live feed sources for hurricanes and cyclones around the world, as well as recent weather radar imagery for the United States.　ArcGIS Online subscription required for hurricane and cyclone layer.

【实验步骤】

1. 新建项目

.NET 框架：与 ArcGIS Runtime SDK 对应
模板列表：Visual C#/Windows/Windows Classic Desktop
模板：ArcGIS Runtime Application（WPF）
名称：UrlWebMapViewer
位置：D：\ArcGISRuntimeTutorial

2. 编译应用程序

详情见第 1 章第 3 节。

3. 修改 MainWindow.xaml 标记语言

在标签 Grid 里，修改 XAML 代码，将网格设置为 2 行。

```
<Grid.RowDefinitions>
    <RowDefinition Height="auto"/>
    <RowDefinition Height="*"/>
</Grid.RowDefinitions>
```

插入工具条：在网格第一行插入工具条。

```
<ToolBar Grid.Row="0">
</ToolBar>
```

加入命令按钮：在工具条加入命令按钮，用于确认用户的底图选择，在点击事件中更改底图。

```
<Button Content="MapViewerUrl" Name="btMapViewerUrl" Click="btMapViewerUrl_Click"/>
```

修改地图视图：更改地图视图在网格中的行号，添加名称。

```
<esri:MapView Grid.Row="1" Name="mapView1" Map="{Binding Map, Source={StaticResource MapViewModel}}"/>
```

4. 修改 MainWindow.xaml.cs 后台代码

进入后台代码：在 MainWindow.xaml 点击右键，选择 View Code，进入 MainWindow.xaml.cs

通过按钮点击事件，更改底图：

```
private async void btMapViewerUrl_Click(object sender, RoutedEventArgs e)
{
    Uri uri = new Uri("https://www.arcgis.com/home/webmap/viewer.html?webmap=69fdcd8e40734712aaec34194d4b988c");
```

```
    Map webmap = await Map.LoadFromUriAsync(uri);
    mapView1.Map = webmap;
}
```

在工具条上另外添加两个按钮，分别利用项目详情和项目数据 JSON 字符串创建地图。

【实验结果】

点击按钮，地图将切换至对应的底图。

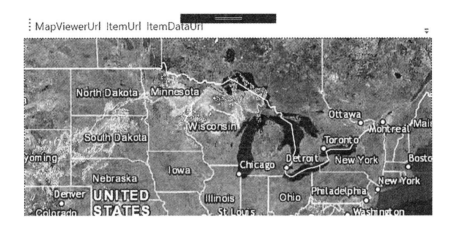

【实验讲解】

(1) MVC 全名是 Model View Controller，是模型（model）-视图（view）-控制器（controller）的缩写，这是一种软件设计典范，用一种业务逻辑、数据、界面显示分离的方法组织代码，将业务逻辑聚集到一个部件里面，在改进和个性化定制界面及用户交互的同时，不需要重新编写业务逻辑。MVC 用于将传统的输入、处理和输出功能映射在一个逻辑的图形化用户界面中。

(2) 最终 MainWindow.xaml 的主要代码如下：

```
<Grid.RowDefinitions>
    <RowDefinition Height = "auto" />
    <RowDefinition Height = " * " />
</Grid.RowDefinitions>
<ToolBar Grid.Row = "0">
    <Button Content = "MapViewerUrl" Name = "btMapViewerUrl" Click = "btMapViewerUrl_Click" />
    <Button Content = "ItemUrl" Name = "btItemUrl" Click = "btItemUrl_Click" />
    <Button Content = "ItemDataUrl" Name = "btItemDataUrl" Click = "btItemDataUrl_Click" />
</ToolBar>
```

```xml
<esri:MapView Grid.Row="1" Name="mapView1" Map="{Binding Map,
Source={StaticResource MapViewModel}}"/>
```

（3）最终 MainWindow.xaml.cs 的主要代码如下：

```csharp
private async void btMapViewerUrl_Click(object sender, RoutedEventArgs e)
{
    Uri uri = new Uri("https://www.arcgis.com/home/webmap/viewer.html?webmap=69fdcd8e40734712aaec34194d4b988c");
    Map webmap = await Map.LoadFromUriAsync(uri);
    mapView1.Map = webmap;
}
private async void btItemUrl_Click(object sender, RoutedEventArgs e)
{
    Uri uri = new Uri("https://www.arcgis.com/home/item.html?id=69fdcd8e40734712aaec34194d4b988c");
    Map webmap = await Map.LoadFromUriAsync(uri);
    mapView1.Map = webmap;
}
private async void btItemDataUrl_Click(object sender, RoutedEventArgs e)
{
    Uri uri = new Uri("https://www.arcgis.com/sharing/rest/content/items/69fdcd8e40734712aaec34194d4b988c/data?");
    Map webmap = await Map.LoadFromUriAsync(uri);
    mapView1.Map = webmap;
}
```

第8节 门户地图浏览器

【实验目的】

编写门户网站网络地图浏览器 PortalViewer，根据门户地图详情标识符 id 设置地图。

【实验数据】

Portal 的内容见下图，其网址为：

https://www.arcgis.com/home/item.html?id=69fdcd8e40734712aaec34194d4b988c

在上述网址中，找到32位的地图详情标识符：

69fdcd8e40734712aaec34194d4b988c

【实验步骤】

1. 新建项目

.NET 框架：与 ArcGIS Runtime SDK 对应
模板列表：Visual C#/Windows/Windows Classic Desktop
模板：ArcGIS Runtime Application（WPF）
名称：PortalViewer
位置：D:\ArcGISRuntimeTutorial

2. 编译应用程序

详情见第 1 章第 3 节。

3. 确认 MainWindow.xaml 标记语言

`<esri:MapView Name="mapView1" />`

4. 修改 MainWindow.xaml.cs 后台代码

进入后台代码：在 MainWindow.xaml 点击右键，选择 View Code，进入 MainWindow.xaml.cs

异步创建 ArcGISPortal 后，通过 id 异步创建 PortalItem，作为底图。

```
public async void Init()
{
    ArcGISPortal arcgisOnline = await ArcGISPortal.CreateAsync();
    PortalItem portalItem = await PortalItem.CreateAsync(arcgisOnline, "69fdcd8e40734712aaec34194d4b988c");
```

```
    Map map = new Map(portalItem);
    mapView1.Map = map;
}
```
　　在构造函数中调用初始化代码进行初始化：
```
public MainWindow()
{
    InitializeComponent();
    Init();
}
```
【实验结果】
　　同本章第 7 节实验：URL 在线网络地图浏览器 UrlWebMapViewer。

第 2 章 常见图层类型

数据被誉为 GIS 的血液。ArcGIS Runtime 支持的数据来源和类型非常丰富，包括在线数据源和离线数据源(本地 SHP 文件、本地地理数据库)等。

在线数据源主要来自 ESRI Portal、Online、Google 地图、Bing 地图、百度地图、高德地图，常用作地图底图，适用于地图浏览、查询和路径分析、地理处理等操作。采用切片表达的栅格数据特别适合地图浏览；而分析功能在服务器端(云端)进行，客户端则只进行渲染。后来引入了矢量切片，则可以进行本地分析和渲染。

支持 SHP 文件和本地轻量级地理数据库，则大大地扩展了 ArcGIS Runtime 对业务的适应能力，这不仅免除了服务器的租用或搭建成本、减少了数据转换操作、提高了系统开发速度、降低了调试难度，而且极大地增加了系统安全性，可以在断网的条件下存储敏感信息和保密数据。

第 1 节 ArcGIS 地图图像图层浏览器

ArcGIS 地图图像图层(ArcGIS map image layer)是来自 ArcGIS 地图服务器(ArcGIS map server)的地图。地图服务可以包含多个图层，服务器每次收到请求后进行渲染并作为一个单独的栅格图像返回给客户端，所以可显示最新的数据。ArcGIS 地图图像服务图层适合显示经常周期性变化的要素，或者需要进行过滤查询的要素。

【实验目的】

通过 URL 浏览在线图像图层，需要网络连接。

【实验数据】

ArcGIS Server sample web service 在线地图：http://sampleserver5.arcgisonline.com/arcgis/rest/services/Elevation/WorldElevations/MapServer

【实验步骤】

1. 新建项目

.NET 框架：与 ArcGIS Runtime SDK 对应
模板列表：Visual C#/Windows/Windows Classic Desktop
模板：ArcGIS Runtime Application (WPF)
名称：ArcGISMapImageLayerViewer
位置：D:\ArcGISRuntimeTutorial

2. 编译项目

见第 1 章第 3 节。

3. 修改 MainWindow.xaml 标记语言

```
<Grid>
    <esri:MapView Name="mapView1">
        <esri:Map/>
    </esri:MapView>
</Grid>
```

4. 修改 MainWindow.xaml.cs 代码

添加命名空间：
```
using Esri.ArcGISRuntime.Mapping;
```

编写 Init 函数，添加 ArcGISMapImageLayer 图层：
```
void Init()
{
    string url = "http://sampleserver5.arcgisonline.com/arcgis/rest/services/Elevation/WorldElevations/MapServer";
    Uri uri = new Uri(url);
    ArcGISMapImageLayer imageLayer = new ArcGISMapImageLayer(uri);
    mapView1.Map.Basemap.BaseLayers.Add(imageLayer);
}
```

在构造函数中调用 Init 函数：
```
public MainWindow()
{
    InitializeComponent();
    Init();
}
```

【实验结果】

第 2 节　ArcGIS 地图图像图层可见性

ArcGIS 地图图像图层（ArcGIS map image layer）中每个子图层的底层服务要素表（service feature table）都可以被访问。这些图层既可以进行属性查询、空间查询、时态查询；也可以进行统计或者关联查找。图层可见性是图层的基本属性。

【实验目的】

通过列表视图控制子图层的可见性。

【实验数据】

ArcGIS Server sample web service 在线地图：

http://sampleserver6.arcgisonline.com/arcgis/rest/services/SampleWorldCities/MapServer

【实验步骤】

1. 新建项目

.NET 框架：与 ArcGIS Runtime SDK 对应

模板列表：Visual C#/Windows/Windows Classic Desktop

模板：ArcGIS Runtime Application（WPF）

名称：ArcGISMapImageLayerSublayers

位置：D:\ArcGISRuntimeTutorial

2. 编译项目

见第 1 章第 3 节。

3. 修改 MainWindow.xaml 标记语言

```
<Grid>
    <esri:MapView Name="mapView1">
        <esri:Map/>
    </esri:MapView>
    <ListView Name="sublayerListView" Opacity="0.5" Margin="10" VerticalAlignment="Top" HorizontalAlignment="Left" Width="100">
        <ListView.ItemTemplate>
            <DataTemplate>
                <CheckBox Content="{Binding Name}" IsChecked="{Binding IsVisible}" Margin="2"/>
            </DataTemplate>
        </ListView.ItemTemplate>
    </ListView>
```

```
</Grid>
```

4. 修改 MainWindow.xaml.cs 代码

添加命名空间：

```
using Esri.ArcGISRuntime.Mapping;
```

编写 Init 函数，添加 ArcGISMapImageLayer 图层：

```
void Init()
{
    string url = " http://sampleserver6.arcgisonline.com/arcgis/rest/services/SampleWorldCities/MapServer";
    Uri uri = new Uri(url);
    ArcGISMapImageLayer imageLayer = new ArcGISMapImageLayer(uri);
    await imageLayer.LoadAsync();
    mapView1.Map.Basemap.BaseLayers.Add(imageLayer);
    sublayerListView.ItemsSource = imageLayer.Sublayers;
}
```

在构造函数中调用 Init 函数：

```
public MainWindow()
{
    InitializeComponent();
    Init();
}
```

【实验结果】

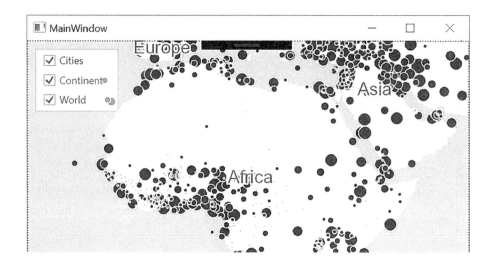

第 3 节　在线 ArcGIS 切片图层浏览器

ArcGIS 切片图层(ArcGIS tiled layer)是为 ArcGIS 服务或切片包(tile package，tpk)提供的栅格切片。

【实验目的】

开发一个 ArcGIS 切片图层在线浏览器 ArcGISTiledLayerViewer。

【实验数据】

ArcGIS Server sample web service 在线地图：
http://services.arcgisonline.com/arcgis/rest/services/World_Topo_Map/MapServer

【实验步骤】

1. 新建项目

.NET 框架：与 ArcGIS Runtime SDK 对应
模板列表：Visual C#/Windows/Windows Classic Desktop
模板：ArcGIS Runtime Application（WPF）
名称：ArcGISTiledLayerViewer
位置：D:\ArcGISRuntimeTutorial

2. 编译应用程序

详情见第 1 章第 3 节。

3. 修改 MainWindow.xaml 标记语言

```
<Grid>
    <esri:MapView Name="mapView1">
        <esri:Map/>
    </esri:MapView>
</Grid>
```

4. 修改 MainWindow.xaml.cs 代码

添加命名空间：
```
using Esri.ArcGISRuntime.Mapping;
```
编写 Init 函数：
```
void Init()
{
    string url = "http://services.arcgisonline.com/arcgis/rest/services/World_Topo_Map/MapServer";
    Uri uri = new Uri(url);
```

```
    ArcGISTiledLayer imageLayer = new ArcGISTiledLayer(uri);
    await imageLayer.LoadAsync();
    mapView1.Map.Basemap.BaseLayers.Add(imageLayer);
}
```
在构造函数中调用 Init 函数：
```
public MainWindow()
{
    InitializeComponent();
    Init();
}
```
【实验结果】

第 4 节　ArcGIS 矢量切片图层浏览器

矢量切片利用新技术来控制可交互的地图展示方式，从而实现在移动端或者浏览器端自定义个性化的地图样式，例如可以根据内容进行智能制图和实时分析，并展示在基础地图上。

矢量切片继承了矢量数据和切片地图的双重优势，具有以下优点。

(1) 灵活：具有更细粒度划分，分要素和专题。

(2) 无损：矢量表达，信息损失量小。

(3) 量小：请求指定地物的信息，直接在客户端获取，无需再次请求服务器。

(4) 高效：数据量小、只进行按需更新。

(5) 定制：矢量切片可以按照用户赋予的样式，在客户端或者服务器端渲染，样式可改变和定制。

(6) 只读：虽然是矢量格式，但不可编辑。如果编辑，可使用 OGC 的 WFS，或者本地 SHP、地理数据库等。

ArcGIS 的矢量切片利用协议缓冲(Protocol Buffers)的紧凑二进制格式来传输，前端通过解析样式动态渲染矢量切片数据。

【实验目的】

开发一个可以从多种矢量切片图层中进行选择的浏览器。

【实验数据】

ArcGIS Server sample web service 在线地图，包括中世纪、彩铅、报纸、新星和世界街道夜间图层，网址分别为：

{"Mid-Century", "http://www.arcgis.com/home/item.html?id=7675d44bb1e4428aa2c30a9b68f97822"},
{"Colored Pencil", "http://www.arcgis.com/home/item.html?id=4cf7e1fb9f254dcda9c8fbadb15cf0f8"},
{"Newspaper", "http://www.arcgis.com/home/item.html?id=dfb04de5f3144a80bc3f9f336228d24a"},
{"Nova", "http://www.arcgis.com/home/item.html?id=75f4dfdff19e445395653121a95a85db"},
{"World Street Map (Night)", "http://www.arcgis.com/home/item.html?id=86f556a2d1fd468181855a35e344567f"}

【实验步骤】

1. 新建项目

.NET 框架：与 ArcGIS Runtime SDK 对应

模板列表：Visual C#/Windows/Windows Classic Desktop

模板：ArcGIS Runtime Application (WPF)

名称：ArcGISVectorTiledLayerViewer

位置：D:\ArcGISRuntimeTutorial

2. 编译应用程序

见第 1 章第 3 节。

3. 修改 MainWindow.xaml 标记语言

```
<Grid>
    <esri:MapView Name="mapView1">
        <esri:Map/>
    </esri:MapView>
    <ComboBox Name="cbLayers" SelectionChanged="cbLayers_SelectionChanged" Opacity="0.5" Margin="10" HorizontalAlignment="Left" VerticalAlignment="Top"/>
</Grid>
```

4. 修改 MainWindow.xaml.cs 代码

添加命名空间：

```
using Esri.ArcGISRuntime.Mapping;
```

存储图层 url：

```
Dictionary<string, string> _dictUrl = new Dictionary<string, string>()
{
    {"Mid-Century", "http://www.arcgis.com/home/item.html?id=7675d44bb1e4428aa2c30a9b68f97822"},
    {"Colored Pencil","http://www.arcgis.com/home/item.html?id=4cf7e1fb9f254dcda9c8fbadb15cf0f8"},
    {"Newspaper", "http://www.arcgis.com/home/item.html?id=dfb04de5f3144a80bc3f9f336228d24a"},
    {"Nova"," http://www.arcgis.com/home/item.html?id=75f4dfdff19e445395653121a95a85db"},
    {"World Street Map (Night)", "http://www.arcgis.com/home/item.html?id=86f556a2d1fd468181855a35e344567f"}
};
```

编写 Init 函数，添加 ArcGISMapImageLayer 图层：

```
void Init()
{
    cbLayers.ItemsSource = _dictUrl.Keys;
    cbLayers.SelectedIndex = 0;
}
```

响应用户图层选择操作：

```
private void cbLayers_SelectionChanged(object sender, SelectionChangedEventArgs e)
{
    string lyrName = e.AddedItems[0].ToString();
    string lyrUrl = _dictUrl[lyrName];
    ArcGISVectorTiledLayer lyr = new ArcGISVectorTiledLayer(new Uri(lyrUrl));
    mapView1.Map.Basemap = new Basemap(lyr);
}
```

在构造函数中调用 Init 函数：

```
public MainWindow()
{
```

```
    InitializeComponent();
    Init();
}
```
【实验结果】

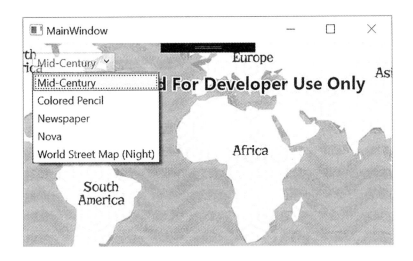

第 5 节　要素集合图层门户浏览器

要素集合可以保存在地图中或者保存为独立的门户项目。如果需要在不同的地图中共享要素集合，最好是保存在用户门户项目中。在编辑门户项目中的要素之后，必须显式保存到原出处。

【实验目的】

开发 FeatureCollectionLayerPortalViewer，读取门户的要素集合图层，添加至业务图层进行显示。

【实验数据】

ArcGIS Server sample web service 在线地图，门户要素 ID 为：
5ffe7733754f44a9af12a489250fe12b

【实验步骤】

1. 新建项目

.NET 框架：与 ArcGIS Runtime SDK 对应

模板列表：Visual C#/Windows/Windows Classic Desktop

模板：ArcGIS Runtime Application（WPF）

名称：FeatureCollectionLayerPortalViewer

位置：D：\ArcGISRuntimeTutorial

2. 编译应用程序

见第 1 章第 3 节。

3. 修改 MainWindow.xaml 标记语言

```
<Grid>
    <esri:MapView Name="mapView1">
        <esri:Map/>
</Grid>
<Grid>
    <esri:MapView Name="mapView1" Map="{Binding Map,Source={StaticResource MapViewModel}}"/>
</Grid>
```

4. 修改 MainWindow.xaml.cs 代码

添加命名空间：

```
using Esri.ArcGISRuntime.Data;
using Esri.ArcGISRuntime.Mapping;
using Esri.ArcGISRuntime.Portal;
```

编写 Init 函数：

```
void Init()
{
    string portalItemId="5ffe7733754f44a9af12a489250fe12b";
    //打开包含要素集合的门户项目
    ArcGISPortal portal = await ArcGISPortal.CreateAsync();
    PortalItem collectionItem = await PortalItem.CreateAsync(portal, portalItemId);
    //验证该项是否为要素集合
    if (collectionItem.Type == PortalItemType.FeatureCollection)
    {
        //从项目创建新的 FeatureColletion
        FeatureCollection featCollection = new FeatureCollection(collectionItem);
        //创建一个图层以显示集合
        FeatureCollectionLayer featCollectionLayer = new FeatureCollectionLayer(featCollection);
        featCollectionLayer.Name = collectionItem.Title;
        mapView1.Map.OperationalLayers.Add(featCollectionLayer);
```

        }
}
　　在构造函数中调用 Init 函数：
```
public MainWindow()
{
    InitializeComponent();
    Init();
}
```
【实验结果】

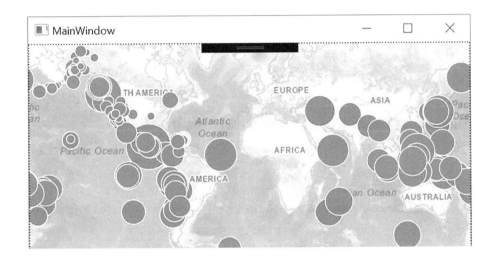

第6节　网络地图服务浏览器

　　Web Map Service（WMS）网络地图服务显示来自网络服务的数据，由服务器端渲染成图像。WMS 是开放地理联盟（Open Geospatial Consortium，OGC）发布的在线地图图像服务标准。WMS 服务使用服务器预定义的符号提供地图，因此不支持自定义渲染或可视化的要素选择。

【实验目的】
　　开发 WMSViewer，根据 URL 加载业务图层。
【实验数据】
　　在线 WMS 服务，网址为：https：//certmapper.cr.usgs.gov/arcgis/services/geology/africa/MapServer/WMSServer? request=GetCapabilities&service=WMS
【实验步骤】
　　1. 新建项目
　　.NET 框架：与 ArcGIS Runtime SDK 对应

模板列表：Visual C#/Windows/Windows Classic Desktop
模板：ArcGIS Runtime Application（WPF）
名称：WMSViewer
位置：D：\ArcGISRuntimeTutorial

2. 编译应用程序

见第 1 章第 3 节。

3. 修改 MainWindow. xaml

```
<esri:MapView Name = "mapView1" />
```

4. 修改 MainWindow. xaml. cs

加入命名空间：

```
using Esri.ArcGISRuntime.Geometry;
using Esri.ArcGISRuntime.Mapping;
```

在 MainWindow 中添加 Init 函数，加载 WMS，作为业务图层：

```
public void Init()
{
    string url = "https://certmapper.cr.usgs.gov/arcgis/services/geology/africa/MapServer/WMSServer? request = GetCapabilities&service=WMS";
    Uri wmsUri = new Uri(url);
    //保留要显示的唯一标识 WMS 图层名称的列表
    List<String> wmsLayerNames = new List<string> { "0" };
    WmsLayer myWmsLayer = new WmsLayer(wmsUri, wmsLayerNames);
    Map myMap = new Map();
    //设置初始视点
    myMap.InitialViewpoint = new Viewpoint(
            new MapPoint(25.450, -4.59, new SpatialReference(4326)),
1000000);
    myMap.OperationalLayers.Add(myWmsLayer);
    mapView1.Map = myMap;
}
```

在 MainWindow 构造函数中添加调用 Init( )

```
public MainWindow()
{
    InitializeComponent();
    Init();
```

}
【实验结果】

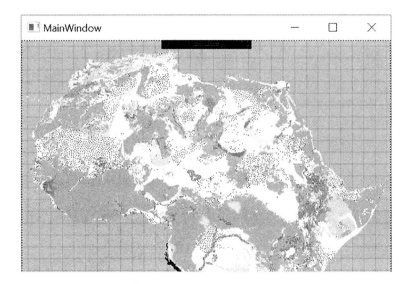

第3章 离线地图

离线地图有两种使用模式：半离线模式和全离线模式。

（1）半离线模式。用户提前规划任务，在联网时下载好目标区域的地图；在断开 internet 连接时查看、收集和更新数据；重新连接后，可以同步地图、发送更新，以及获取地图更新。

（2）全离线模式。设备在整个作业过程中都不联网。数据通过本地桌面软件做好后，直接部署到移动端；用户可以浏览和编辑本地文件或地理数据库，从而获得桌面端的顺畅体验。

离线地图可以把桌面端渲染好的地图拿来即用，无需二次渲染；支持打包矢量切片，解决地图包过大的问题；支持路径分析、地理编码和本地编辑，具有安全、稳定、开发效率高和响应速度快等众多优点。

离线数据是离线地图的重要准备工作。在学完本章后，读者可以动手使用 ArcGIS Desktop 或者 ArcGIS Pro 对数据进行优化后，制作移动地图包（MMPK）、地图包（MPK）、图层包（LPK）、SHP 等，然后加载到开发的 App 中，在学完后面的章节后，进一步添加浏览、查询、分析和编辑功能。

移动数据包优化的技巧很多，例如，针对矢量数据去掉多余的属性、去掉冗余的字段、统一坐标系和投影、只保留必要的精度（例如调查类数据的坐标精度一般只需要保留到整数米），栅格数据保留适当的深度和位数。

第1节 移动地图包浏览器

移动地图包（mobile map package）是以".mmpk"结尾的单独文件，它可以将你组织的 maps、资源、道路网和坐标集成到一个文件。

【实验目的】

（1）熟悉按钮 Button 的点击事件。

（2）开发 MmpkViewer，打开并浏览本地移动地图包。

【实验数据】

D:\ArcGISRuntimeTutorial\ArcGISRuntimeSampleData\Yellowstone\Yellowstone.mmpk

【实验步骤】

1. 新建项目

.NET 框架：与 ArcGIS Runtime SDK 对应

模板列表：Visual C#/Windows/Windows Classic Desktop
模板：ArcGIS Runtime Application（WPF）
名称：MmpkViewer
位置：D：\ArcGISRuntimeTutorial

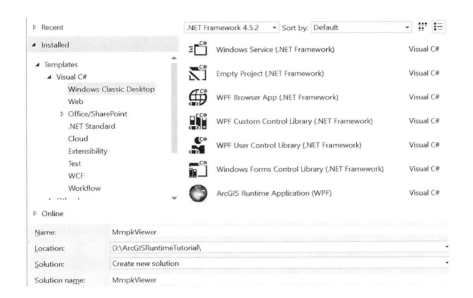

2. 编译项目

见第 1 章第 3 节。

3. 修改 MainWindow. xaml 标记语言

```
<Grid.RowDefinitions>
    <RowDefinition Height = "auto" />
    <RowDefinition Height = " * " />
</Grid.RowDefinitions>
<ToolBar Grid.Row = "0">
    <Button Content = "OpenMmpk" Name = "btOpenMmpk" Click = "btOpenMmpk_Click" />
</ToolBar>
<esri:MapView Grid.Row = "1" Name = "mapView1" />
```

4. 修改 MainWindow. xaml. cs 代码

```
private async void btOpenMmpk_Click(object sender, RoutedEventArgs e)
{
```

```
    string mmpkPath = @ "D:\ArcGISRuntimeTutorial\ArcGISRuntimeSampleData\Yellowstone\Yellowstone.mmpk";
    MobileMapPackage mmpk = await MobileMapPackage.OpenAsync(mmpkPath);
    if (mmpk.Maps.Count() > 0)
        mapView1.Map = mmpk.Maps.First();
}
```

【实验结果】

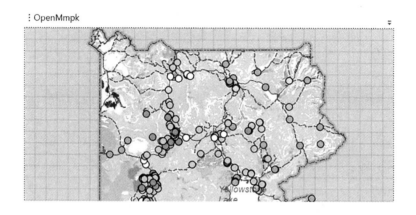

【实验分析】

地图数据包格式是面向手机设备的优化格式,因为压缩存储、小巧而快速、容易分享,相比于传统的地图包(tpk、vtpk 等),它保存所有的 feature 要素。

mmpk 可以将各种地图和数据资源打包,其中也包括矢量切片数据,可轻松部署到终端使用。数据是存储于压缩的 mobile GDB 中的,底图上展示的内容是要素,可供查询和分析,同时支持路径规划、地址编码等高级分析应用。移动地图包既解决了移动端符号渲染问题,又结合矢量切片底图解决了地图包过大的问题,支持离线查询、分析等各种应用场景。

第 2 节　矢量图层地理包浏览器

存储在地理包(.gpkg)中的要素表格使用要素图层进行显示。地理包标准开放,平台独立,便携,自描述,符合开放地理联盟(OGC)规范,使用 SQLite 数据库存储。地理包要素能够进行编辑和保存。

【实验目的】

读取本地地理包,将矢量数据显示在业务图层中。

【实验数据】

本地地理包(*.gpkg):
AuroraCO_gpkg\\AuroraCO.gpkg

第 2 节 矢量图层地理包浏览器

【实验步骤】

1. 新建项目

.NET 框架：与 ArcGIS Runtime SDK 对应
模板列表：Visual C#/Windows/Windows Classic Desktop
模板：ArcGIS Runtime Application（WPF）
名称：GeoPackageViewer
位置：D：\ArcGISRuntimeTutorial

2. 编译项目

详情见第 1 章第 3 节。

3. 修改 MainWindow.xaml 标记语言

添加名称属性：

```
<Grid>
    <esri:MapView Name = "mapView1" Map = "{Binding Map, Source = {StaticResource MapViewModel}}"/>
</Grid>
```

4. 修改 MainWindow.xaml.cs 代码

添加命名空间：

```
using Esri.ArcGISRuntime.Data;
using Esri.ArcGISRuntime.Mapping;
```

编写 Initialize 函数，逻辑关系为 GeoPackage-> FeatureTable-> FeatureLayer。

```
private async void Initialize()
{
    string gpkg = "AuroraCO_gpkg\\AuroraCO.gpkg";
    string geoPackagePath = System.IO.Path.Combine(AppDomain.CurrentDomain.BaseDirectory, gpkg);
    //Open the GeoPackage.
    GeoPackage geoPackage = await GeoPackage.OpenAsync(geoPackagePath);
    //Read the feature tables and get the first one.
    FeatureTable geoPackageTable = geoPackage.GeoPackageFeatureTables.FirstOrDefault();
    if (geoPackageTable == null) return;
    //Create a layer to show the feature table.
    FeatureLayer newLayer = new FeatureLayer(geoPackageTable);
```

```
        await newLayer.LoadAsync();
        await mapView1.SetViewpointAsync(new Viewpoint(newLayer.FullEx-
tent));
        //Add the feature table as a layer to the map (with default symbolo-
gy).
        mapView1.Map.OperationalLayers.Add(newLayer);
}
```
在构造函数中调用 Init 函数：
```
public MainWindow()
{
    InitializeComponent();
    Initialize ();
}
```
【实验结果】

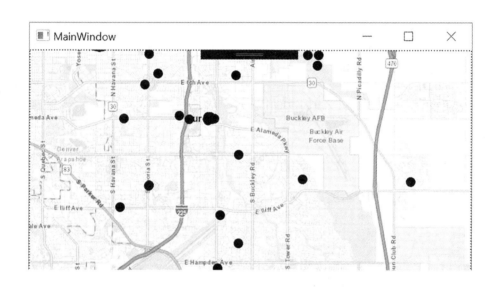

第 3 节　栅格图层地理包浏览器

存储在地理包(.gpkg)中的栅格数据使用栅格图层进行显示。地理包符合开放地理联盟(OGC)规范。

【实验目的】

读取本地地理包，将栅格显示在业务图层中。

【实验数据】

本地地理包(*.gpkg)：
AuroraCO_gpkg\\AuroraCO.gpkg

第 3 节　栅格图层地理包浏览器

【实验步骤】

1. 新建项目

.NET 框架：与 ArcGIS Runtime SDK 对应
模板列表：Visual C#/Windows/Windows Classic Desktop
模板：ArcGIS Runtime Application（WPF）
名称：RasterLayerGeoPackageViewer
位置：D：\ArcGISRuntimeTutorial

2. 编译项目

详情见第 1 章第 3 节。

3. 修改 MainWindow.xaml 标记语言

对地图视图添加名称属性：

```
<esri:MapView Name="mapView1" Map="{Binding Map, Source={Static ResourceMapViewModel}}" />
```

设计 UI：
StackPanel：容器
TextBox：本地 gpkg 路径
Button：加载 gpkg

```
...Viewer\RasterLayerGeoPackageViewer\bin\Debug\AuroraCO_gpkg\AuroraCO.gpkg
Load
```

修改 xaml 代码如下：

```xml
<StackPanel VerticalAlignment="Top">
    <TextBox Name="tbGpkPath" />
    <Button Name="btLoad" Content="Load" Click="btLoad_Click" />
</StackPanel>
```

4. 修改 MainWindow.xaml.cs 代码

添加命名空间：

```csharp
using Esri.ArcGISRuntime.Data;
using Esri.ArcGISRuntime.Mapping;
using Esri.ArcGISRuntime.Rasters;
```

初始化地图，填充默认 gpkg 路径：

```csharp
public MainWindow()
{
    InitializeComponent();
```

```csharp
    InitMap();
}
string _gpkg = "AuroraCO_gpkg\\AuroraCO.gpkg";
void InitMap()
{
    tbGpkPath.Text = Path.Combine(AppDomain.CurrentDomain.BaseDirectory, _gpkg);
}
```

打开当地 gpkg，获取第一个栅格，添加进业务图层中显示，逻辑关系为 GeoPackage→Raster→RasterLayer。

```csharp
private async void btLoad_Click(object sender, RoutedEventArgs e)
{
    //Open the GeoPackage.
    GeoPackage myGeoPackage = await GeoPackage.OpenAsync(tbGpkPath.Text);
    //Read the raster images and get the first one.
    Raster raster = myGeoPackage.GeoPackageRasters.FirstOrDefault();
    //Make sure an image was found.
    if (raster == null) return;
    //Create a layer to show the raster.
    RasterLayer rl = new RasterLayer(raster);
    await rl.LoadAsync();
    await mapView1.SetViewpointAsync(new Viewpoint(rl.FullExtent));
    //Add the image as a raster layer to the map.
    mapView1.Map.OperationalLayers.Add(rl);
}
```

【实验结果】

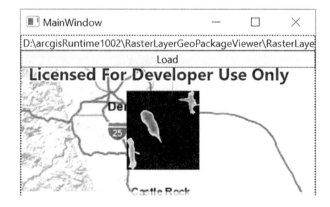

第 4 节 TPK 浏览器

切片包(tile package，TPK)也称为切片缓存，在地图应用中能作为 Web 切片发布(图像)，常用作底图。

将切片打包成单个压缩的切片包文件，可以通过网络共享，配置离线运行，从而脱离 ArcGIS Server 和 ArcGIS Online 独立运行。

TPK 是一个 zip 格式的压缩包，把 TPK 文件后缀名改成.zip 进行解压，就会看到目录结构。

切片所生成的底图不必等待创建图像，显示速度通常比其他类型的底图显示更快。尽管创建切片包会花费一些时间，但其成本属于一次性费用。

【实验目的】

打开并浏览创建的本地 TPK。

【实验数据】

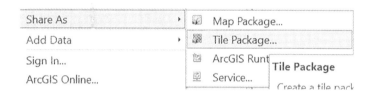

【实验步骤】

1. 新建项目

.NET 框架：与 ArcGIS Runtime SDK 对应
模板列表：Visual C#/Windows/Windows Classic Desktop
模板：ArcGIS Runtime Application（WPF）
名称：TpkViewer
位置：D：\ArcGISRuntimeTutorial

2. 编译项目

详情见第 1 章第 3 节。

3. 修改 MainWindow.xaml 标记语言

对地图视图添加名称属性：
`<esri:MapView Name="mapView1" Map="{Binding Map,Source={Static Resource MapViewModel}}" />`

设计 UI，添加 TPK 文件位置文本框和 TPK 加载按钮：
`<StackPanel Background="Wheat" Opacity="0.9" Margin="3" Orienta-`

```xml
tion="Horizontal" HorizontalAlignment="Left" VerticalAlignment="Top">
    <TextBox Name="tbTpk" Text="SanFrancisco\SanFrancisco.tpk" Margin="3" Padding="3" MinWidth="200"/>
    <Button Name="btLoadTpk" Content="LoadTpk" Padding="3" Margin="3" Click="btLoadTpk_Click"/>
</StackPanel>
```

4. 修改 MainWindow.xaml.cs 代码

添加命名空间：
```csharp
using Esri.ArcGISRuntime.Mapping;
using System.IO;
```

响应 TPK 加载事件：
```csharp
async void btLoadTpk_Click(object sender, RoutedEventArgs e)
{
    string tpkPath = Path.Combine(AppDomain.CurrentDomain.BaseDirectory, tbTpk.Text.Trim());
    //Load the saved tile cache.
    TileCache cache = new TileCache(tpkPath);
    //Load the cache.
    await cache.LoadAsync();
    //Create a tile layer with the cache.
    ArcGISTiledLayer tiledLayer = new ArcGISTiledLayer(cache);
    mapView1.Map = new Map(new Basemap(tiledLayer));
}
```

【实验结果】

第 5 节　矢量切片包浏览器

矢量切片包(vector tile package，VTPK)存储矢量切片，包含跨尺度的矢量数据。区别于栅格切片，矢量切片能够在地图缩放时适应显示设备的屏幕分辨率。

【实验目的】

开发 VtpkViewer。

【实验数据】

(1)可以在 ArcGIS Pro 中通过导出矢量切片包(vector tile package，VTPK)制作；
(2)也可以下载别人已经制作好的 VTPK。

【实验步骤】

1. 新建项目

.NET 框架：与 ArcGIS Runtime SDK 对应
模板列表：Visual C#/Windows/Windows Classic Desktop
模板：ArcGIS Runtime Application (WPF)
名称：VtpkViewer
位置：D：\ArcGISRuntimeTutorial

2. 编译项目

详情见第 1 章第 3 节。

3. 修改 MainWindow. xaml 标记语言

对地图视图添加名称属性：
<esri:MapView Name="mapView1" Map="{Binding Map, Source={StaticResource MapViewModel}}" />

设计 UI，添加堆叠面板 StackPanel，子控件包括文本框和按钮。
文本框：输入 vtpk 路径。
按钮：加载 vtpk。

4. 修改 MainWindow. xaml. cs 代码

添加命名空间：
using Esri.ArcGISRuntime.Mapping;

加载 vtpk 图层，逻辑关系为 vtpkPath → VectorTileCache → LoadAsync → ArcGISVectorTiledLayer：

```
private async void btLoadVtpk_Click(object sender, RoutedEventArgs e)
{
    string vtpkPath = tbVtpk.Text.Trim();
    if (! File.Exists(vtpkPath))
       {
           MessageBox.Show( $"{vtpkPath} does not exists!");
           return;
       }
    //Load the vector tile cache.
    VectorTileCache cache = new VectorTileCache(vtpkPath);
    //Load the cache.
    await cache.LoadAsync();
    //Create a vector tile layer with the cache.
     ArcGISVectorTiledLayer tiledLayer = new ArcGISVectorTiledLayer(cache);
    mapView1.Map = new Map(new Basemap(tiledLayer));
}
```

【实验结果】

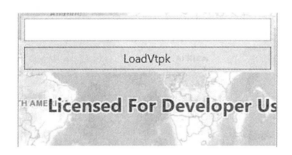

第 6 节　SHP 浏览器

ESRI Shapefile(SHP)是美国环境系统研究所(ESRI)开发的空间数据格式，在 20 世纪 90 年代初的 ArcView 中被首次应用。该文件格式已经成为地理信息软件界的开放标准，可以自由使用而无需支付版权或专利费用，因此绝大部分开放程序或商业软件都可以读取 Shapefile，使它成为重要的交换格式。

由于历史久远，shpfile 也存在一些问题，例如不存储地理数据的拓扑信息，同时对字段长度有限制、兼容性差、容易出现乱码等。但经验丰富的 GISer 有对应的办法解决这些问题。

【实验目的】

开发 ShpViewer，加载并显示本地 SHP 文件。

【实验数据】

本地 shp 文件为小比例尺的中国省级行政区划边界：ChinaProvince \ bou2_4p.shp

【实验步骤】

1. 新建项目

.NET 框架：与 ArcGIS Runtime SDK 对应

模板列表：Visual C#/Windows/Windows Classic Desktop

模板：ArcGIS Runtime Application（WPF）

名称：ShpViewer

位置：D：\ArcGISRuntimeTutorial

2. 编译项目

详情见第 1 章第 3 节。

3. 修改 MainWindow.xaml 标记语言，添加名称属性：

```
<Grid>
    <esri:MapView Name = "mapView1" Map = "{Binding Map, Source = {StaticResource MapViewModel}}" />
</Grid>
```

4. 修改 MainWindow.xaml.cs 代码

添加命名空间：

```
using Esri.ArcGISRuntime.Data;
using Esri.ArcGISRuntime.Mapping;
using System.IO;
```

生成全局变量，用于存储 shp 文件路径：

```
string _shpPath = @"ChinaProvince\bou2_4p.shp";
```

编写 Initialize 函数，逻辑关系为 shapefile→ShapefileFeatureTable→FeatureLayer：

```
void Initialize()
{
    //获取 shp 文件路径。
    string filepath = System.IO.Path.Combine(AppDomain.CurrentDomain.BaseDirectory, _shpPath);
```

```
        //打开shp文件。
    ShapefileFeatureTable myShapefile = await ShapefileFeatureTable.OpenAsync(filepath);
        //生成要素图层用于显示shp文件。
    FeatureLayer newFeatureLayer = new FeatureLayer(myShapefile);
        //添加要素图层至地图。
    mapView1.Map.OperationalLayers.Add(newFeatureLayer);
        //缩放至要素图层。
    await mapView1.SetViewpointGeometryAsync(newFeatureLayer.FullExtent);
}
```

在构造函数中调用Initialize函数：
```
public MainWindow()
{
    InitializeComponent();
    Initialize ();
}
```

第7节　栅格文件浏览器

栅格文件数据源作为栅格图层显示。本地栅格文件最常用的格式为*.tif。

【实验目的】

在全球地图背景下显示北部湾大学周围的真彩色高分辨率卫星遥感影像。

【实验数据】

北部湾大学卫星遥感影像：raster \ 北部湾大学.tif。

【实验步骤】

1. 新建项目

.NET框架：与ArcGIS Runtime SDK对应

模板列表：Visual C#/Windows/Windows Classic Desktop

模板：ArcGIS Runtime Application（WPF）

名称：RasterFileViwer

位置：D：\ArcGISRuntimeTutorial

2. 编译项目

详情见第1章第3节。

3. 修改 MainWindow.xaml 标记语言

对地图视图添加名称属性：

```
<esri:MapView Name="mapView1" Map="{Binding Map,Source={StaticResource MapViewModel}}"/>
```

在地图视图后添加工具条和子控件，设计 UI：

```
<ToolBar Opacity="0.8" HorizontalAlignment="Left" VerticalAlignment="Top" Margin="10">
    <Button Name="btnAddRaster" Content="AddRaster" Click="btnAddRaster_Click"/>
    <Button Name="btnRemoveRaster" Content="RemoveRaster" Click="btnRemoveRaster_Click"/>
</ToolBar>
```

4. 修改 MainWindow.xaml.cs 代码

添加命名空间：

```
using Esri.ArcGISRuntime.Mapping;
using Esri.ArcGISRuntime.Rasters;
```

定义全局变量，存储栅格文件相对路径：

```
string _rasterPath = @"raster\北部湾大学.tif";
```

编写 Initialize 函数，打开栅格文件，设置视点并加载到地图业务图层中进行显示：

```
async void Initialize()
{
    //Get the file name.
    String filepath = System.IO.Path.Combine(AppDomain.CurrentDomain.BaseDirectory, _rasterPath);
    //Load the raster file.
    Raster raster = new Raster(filepath);
    //Create the layer.
    RasterLayer rasterLayer = new RasterLayer(raster);
    //Wait for the layer to load.
    await rasterLayer.LoadAsync();
    rasterLayer.Opacity = 0.6;
    //Set the viewpoint.
    mapView1.Map.InitialViewpoint = new Viewpoint(rasterLayer.FullExtent);
    mapView1.SetViewpoint(new Viewpoint(rasterLayer.FullExtent));
```

```
    //Add the layer to the map.
    mapView1.Map.OperationalLayers.Add(rasterLayer);
}
```
响应添加栅格按钮点击事件，调用 Initialize 函数：
```
private void btnAddRaster_Click(object sender, RoutedEventArgs e)
{
    Initialize();
}
```
响应删除栅格按钮点击事件，移除业务图层：
```
private void btnRemoveRaster_Click(object sender, RoutedEventArgs e)
{
    mapView1.Map.OperationalLayers.Clear();
}
```

【实验结果】

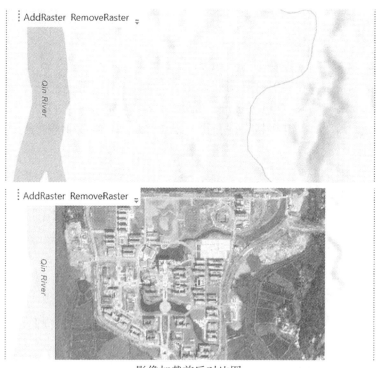

影像加载前后对比图

第 8 节　栅格地理包浏览器

GeoPackage(＊.gpkg)是一个开放的地理空间信息的格式，基于标准、平台独立，可移植、自描述、格式紧凑，适合于移动设备如智能手机等，遵循 OGC 标准，使用单独 SQLite 数据库文件。GeoPackage 文件必须具有扩展名 .gpkg 才能被 ArcGIS 识别。在 ArcGIS Pro 中可使用 SQLite 数据库工具创建 GeoPackage。

【实验目的】

在全球地图背景上加载并显示 AuroraCO.gpkg 地理包。

【实验数据】

AuroraCO 地区本地地理包：AuroraCO_gpkg \ AuroraCO.gpkg

【实验步骤】

1. 新建项目

.NET 框架：与 ArcGIS Runtime SDK 对应

模板列表：Visual C#/Windows/Windows Classic Desktop

模板：ArcGIS Runtime Application（WPF）

名称：RasterGpkgViewer

位置：D：\ArcGISRuntimeTutorial

2. 编译项目

详情见第 1 章第 3 节。

3. 修改 MainWindow.xaml 标记语言

对地图视图添加名称属性：

```
<esri:MapView Name="mapView1" Map="{Binding Map,Source={Static Resource MapViewModel}}"/>
```

在地图视图后添加工具条和子控件，设计 UI：

```
<ToolBar Opacity="0.8" HorizontalAlignment="Left" VerticalAlignment="Top"Margin="10">
    <Button Name="btnAddRaster" Content="AddRaster" Click="btnAddRaster_Click"/>
    <Button Name="btnRemoveRaster" Content="RemoveRaster" Click="btnRemoveRaster_Click"/>
</ToolBar>
```

4. 修改 MainWindow.xaml.cs 代码

添加全局变量，存储 gpkg 相对路径：

```
string _gpkgPath = @"AuroraCO_gpkg\AuroraCO.gpkg";
```

添加命名空间：

```csharp
using Esri.ArcGISRuntime.Data;
using Esri.ArcGISRuntime.Mapping;
using Esri.ArcGISRuntime.Rasters;
```

编写 Initialize 函数,打开栅格文件,设置视点并加载到地图业务图层中进行显示:

```csharp
async void Initialize()
{
    //Get the full path.
    string geoPackagePath=System.IO.Path.Combine(AppDomain.CurrentDomain.BaseDirectory, _gpkgPath);
    //Open the GeoPackage.
    GeoPackage myGeoPackage = await GeoPackage.OpenAsync(geoPackagePath);
    //Read the raster images and get the first one.
    Raster gpkgRaster = myGeoPackage.GeoPackageRasters.FirstOrDefault();
    if (gpkgRaster == null) return;
    //Create a layer to show the raster.
    RasterLayer rl = new RasterLayer(gpkgRaster);
    rl.Opacity = 0.8;
    await rl.LoadAsync();
    mapView1.SetViewpoint(new Viewpoint(rl.FullExtent));
    // Add the image as a raster layer to the map (with default symbology).
    mapView1.Map.OperationalLayers.Add(rl);
}
```

响应添加栅格按钮点击事件,调用 Initialize 函数:

```csharp
private void btnAddRaster_Click(object sender, RoutedEventArgs e)
{
    Initialize();
}
```

响应删除栅格按钮点击事件,移除业务图层:

```csharp
private void btnRemoveRaster_Click(object sender, RoutedEventArgs e)
{
    mapView1.Map.OperationalLayers.Clear();
}
```

【实验结果】

第 9 节　影像服务栅格浏览器

栅格数据常用于地图底图，这些数据常来自遥感影像，在分类、变化分析中得到了广泛应用。栅格值可表示离散值，如土地利用代码；也可表示连续值，例如高程。

【实验目的】

在全球地图背景上加载并显示在线地图。

【实验数据】

ArcGIS online 在线地图栅格服务：

http：//sampleserver6.arcgisonline.com/arcgis/rest/services/NLCDLandCover2001/ImageServer

【实验步骤】

1. 新建项目

.NET 框架：与 ArcGIS Runtime SDK 对应

模板列表：Visual C#/Windows/Windows Classic Desktop

模板：ArcGIS Runtime Application（WPF）

名称：ImageServiceRasterViewer

位置：D：\ArcGISRuntimeTutorial

2. 编译项目

详情见第 1 章第 3 节。

3. 修改 MainWindow.xaml 标记语言

对地图视图添加名称属性：

```xml
<esri:MapView Name="mapView1" Map="{Binding Map, Source={Static
Resource MapViewModel}}" />
```
在地图视图后添加工具条和子控件,设计 UI:
```xml
<ToolBar Opacity="0.8" HorizontalAlignment="Left" Vertical
Alignment="Top" Margin="10">
    <Button Name="btnAddRaster" Content="AddRaster" Click="btnAdd
Raster_Click" />
    <Button Name="btnRemoveRaster" Content="RemoveRaster" Click=
"btnRemoveRaster_Click" />
</ToolBar>
```

4. 修改 MainWindow.xaml.cs 代码

添加命名空间:
```csharp
using Esri.ArcGISRuntime.ArcGISServices;
using Esri.ArcGISRuntime.Mapping;
using Esri.ArcGISRuntime.Rasters;
```
添加全部变量,存储影像服务栅格 url:
```csharp
string _url = "http://sampleserver6.arcgisonline.com/arcgis/rest/services/NLCDLandCover2001/ImageServer";
```
编写 Initialize 函数:
```csharp
async void Initialize()
{
    //Create a Uri to the image service raster.
    var uri = new Uri(_url);
    //Create new image service raster from the Uri.
    ImageServiceRaster imageServiceRaster = new ImageServiceRaster(uri);
    //Load the image service raster.
    await imageServiceRaster.LoadAsync();
    //Create a new raster layer from the image service raster.
    RasterLayer rasterLayer = new RasterLayer(imageServiceRaster);
    //Add the raster layer to the maps layer collection.
    mapView1.Map.OperationalLayers.Add(rasterLayer);
    //Get the service information (metadata) about the image service raster.
    ArcGISImageServiceInfo isi = imageServiceRaster.ServiceInfo;
```

```
    //Zoom the map to the extent of the image service raster (which also
the extent of the raster layer).
    await mapView1.SetViewpointGeometryAsync(isi.FullExtent);
}
```
    响应添加栅格按钮点击事件，调用 Initialize 函数：
```
private void btnAddRaster_Click(object sender, RoutedEventArgs e)
{
    Initialize();
}
```
    响应删除栅格按钮点击事件，移除业务图层：
```
private void btnRemoveRaster_Click(object sender, RoutedEventArgs e)
{
    mapView1.Map.OperationalLayers.Clear();
}
```
【实验结果】

第 10 节　多光谱影像彩色渲染器

栅格数据可以存储单波段信息，如数字高程模型（digital elevation model，DEM），也可以存储多波段信息，例如卫星遥感影像有多个波段，代表了不同波长的电磁波谱上的信息。Landsat7 有 7 个波段，包含从可见光到红外波谱。对多波段进行不同的彩色合成，在植被监测、烟火识别、火灾分析中有特殊的应用。

【实验目的】

Landsat 卫星多光谱影像，裁剪范围为山区，进行真彩色和加彩色合成，以进行活动着火点和燃后范围分析。

【实验数据】

Landsat 卫星影像：landsat \ p114r075_fire.tif

【实验步骤】

1. 新建项目

.NET 框架：与 ArcGIS Runtime SDK 对应
模板列表：Visual C#/Windows/Windows Classic Desktop
模板：ArcGIS Runtime Application（WPF）
名称：MultispectralColorRender
位置：D：\ArcGISRuntimeTutorial

2. 编译项目

详情见第 1 章第 3 节。

3. 修改 MainWindow.xaml 标记语言

对地图视图添加名称属性：

<esri:MapView Name="mapView1" Map="{Binding Map, Source={Static Resource MapViewModel}}" />

设计 UI 界面如下：

在地图视图后添加工具条和子控件：

使用网格（Grid）进行布局，4 行×2 列。

根据 UI 设置编写 XAML 代码：

```
<Grid Margin="10" Background="White" Opacity="0.9" HorizontalAlignment="Left" VerticalAlignment="Top">
    <Grid.RowDefinitions>
        <RowDefinition/>
        <RowDefinition/>
        <RowDefinition/>
        <RowDefinition/>
    </Grid.RowDefinitions>
    <Grid.ColumnDefinitions>
```

```xml
            <ColumnDefinition/>
            <ColumnDefinition/>
        </Grid.ColumnDefinitions>
        <TextBlock Text="Blue" HorizontalAlignment="Right" Margin="3" Grid.Row="0" Grid.Column="0"/>
        <TextBlock Text="Green" HorizontalAlignment="Right" Margin="3" Grid.Row="1" Grid.Column="0"/>
        <TextBlock Text="Red" HorizontalAlignment="Right" Margin="3" Grid.Row="2" Grid.Column="0"/>
        <ComboBox Name="cbBlue" Margin="3" Grid.Row="0" Grid.Column="1"/>
        <ComboBox Name="cbGreen" Margin="3" Grid.Row="1" Grid.Column="1"/>
        <ComboBox Name="cbRed" Margin="3" Grid.Row="2" Grid.Column="1"/>
        <Button Name="btnApply" Content="Apply" Click="btnApply_Click" Grid.Row="3" Grid.ColumnSpan="2"/>
</Grid>
```

4. 修改 MainWindow.xaml.cs 代码

添加命名空间：

```csharp
using Esri.ArcGISRuntime.Mapping;
using Esri.ArcGISRuntime.Rasters;
```

添加全部变量，存储栅格路径和栅格图层：

```csharp
string _rasterPath = @"landsat\p114r075_fire.tif";
RasterLayer _rasterLayer;
```

编写 InitMap 函数，打开栅格数据，设置视点，添加到地图中显示：

```csharp
async void InitMap()
{
    string rasterFile = System.IO.Path.Combine(AppDomain.CurrentDomain.BaseDirectory, _rasterPath);
    Raster raster = new Raster(rasterFile);
    //Create a new raster layer to show the image.
    _rasterLayer = new RasterLayer(raster);
    await _rasterLayer.LoadAsync();
    //Set the initial viewpoint with the raster's full extent.
```

```csharp
    mapView1.Map.InitialViewpoint = new Viewpoint(_rasterLayer.FullExtent);
    //Add the layer to the map.
    mapView1.Map.OperationalLayers.Add(_rasterLayer);
}
```

对 UI 界面初始化：

```csharp
void InitUI()
{
    int nbands = 5; //band count
    IEnumerable<int> bands = Enumerable.Range(0, nbands);
    cbBlue.ItemsSource = bands;
    cbGreen.ItemsSource = bands;
    cbRed.ItemsSource = bands;
    cbBlue.SelectedItem = 0;
    cbGreen.SelectedItem = 1;
    cbRed.SelectedItem = 2;
}
```

根据选择波段，进行彩色合成：

```csharp
private void btnApply_Click(object sender, RoutedEventArgs e)
{
    int red = (int)cbRed.SelectedValue; //red band.
    int green = (int)cbGreen.SelectedValue; //green band.
    int blue = (int)cbBlue.SelectedValue; //blue band.
    int[] bands = { red, green, blue };//band 组合.
    StretchParameters stretchParameters = new StandardDeviationStretchParameters(2); //拉伸方式.
    RgbRenderer rasterRenderer = new RgbRenderer(stretchParameters, bands, null, true);
    _rasterLayer.Renderer = rasterRenderer;
}
```

最后在构造函数中调用地图初始化和界面初始化：

```csharp
public MainWindow()
{
    InitializeComponent();
    InitMap();
    InitUI();
}
```

【实验结果】

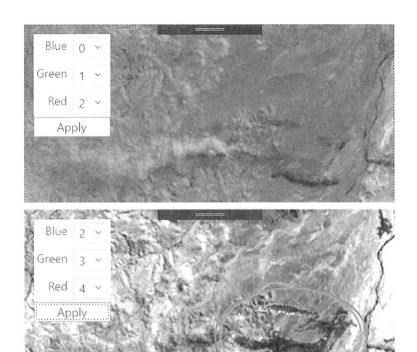

真彩色和标准假彩色合成对比效果图

第 11 节　栅格拉伸渲染器

图像拉伸可以对图像进行增强。拉伸方式参数基类为 Esri.ArcGISRuntime.Rasters.StretchParameters，派生类如下：

Esri.ArcGISRuntime.Rasters.HistogramEqualizationStretchParameters

Esri.ArcGISRuntime.Rasters.MinMaxStretchParameters

Esri.ArcGISRuntime.Rasters.PercentClipStretchParameters

Esri.ArcGISRuntime.Rasters.StandardDeviationStretchParameters

【实验目的】

对栅格图层进行红绿蓝彩色渲染，用户可以调整多光谱影像的各波段。本实验处理三种图像拉伸方式。

【实验数据】

ArcGIS online 在线地图栅格服务。

【实验步骤】

第3章 离线地图

1. 新建项目

.NET 框架：与 ArcGIS Runtime SDK 对应
模板列表：Visual C#/Windows/Windows Classic Desktop
模板：ArcGIS Runtime Application（WPF）
名称：RasterStretchRender
位置：D：\ArcGISRuntimeTutorial

2. 编译项目

详情见第 1 章第 3 节。

3. 修改 MainWindow.xaml 标记语言。

对地图视图添加名称属性：
<esri:MapView Name = " mapView1 " Map = " {Binding Map, Source = {StaticResource MapViewModel}}" />

设计 UI 界面如下：

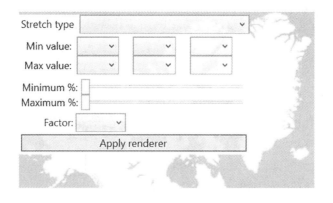

在地图视图后添加工具条和子控件：
为了简化布局，用 StackPanel 简化布局，里面包含 5 行子元素。
第一行：拉伸类型，封装在 StackPanel 里。
第二行：MinMax 拉伸参数，封装在 Grid 里。
第三行：PercentClip 拉伸参数，封装在 Grid 里。
第四行：Standard Deviation 拉伸参数，封装在 Grid 里。
第五行：应用按钮。
根据 UI 设置编写 XAML 代码：

```
< StackPanel Opacity = " 0.8 " Margin = " 10 " Background = " White " HorizontalAlignment="Left" VerticalAlignment="Top">
    <StackPanel Orientation="Horizontal">
```

```xml
<TextBlock Text="Stretch type " HorizontalAlignment="Right" VerticalAlignment="Center" Margin="3"/>
    <ComboBox Name="StretchTypeComboBox" MinWidth="200" SelectionChanged="StretchTypeComboBox_SelectionChanged"/>
</StackPanel>
<Grid Name="MinMaxParamsGrid" Margin="3" Visibility="Visible">
    <Grid.RowDefinitions>
        <RowDefinition />
        <RowDefinition />
    </Grid.RowDefinitions>
    <Grid.ColumnDefinitions>
        <ColumnDefinition />
        <ColumnDefinition />
        <ColumnDefinition />
        <ColumnDefinition />
    </Grid.ColumnDefinitions>
    <TextBlock Grid.Row="0" Grid.Column="0" Text="Min value: " HorizontalAlignment="Right" VerticalAlignment="Center" Margin="3"/>
    <ComboBox Name="MinRedComboBox" Width="50" Grid.Row="0" Grid.Column="1" Foreground="red" HorizontalAlignment="Left" VerticalAlignment="Center" HorizontalContentAlignment="Right"/>
    <ComboBox Name="MinGreenComboBox" Grid.Row="0" Grid.Column="2" Foreground="Green" Width="50" HorizontalAlignment="Left" VerticalAlignment="Center" HorizontalContentAlignment="Right"/>
    <ComboBox Name="MinBlueComboBox" Grid.Row="0" Grid.Column="3" Foreground="Blue" Width="50" HorizontalAlignment="Left" VerticalAlignment="Center" HorizontalContentAlignment="Right"/>
    <TextBlock Grid.Row="1" Grid.Column="0" Text="Max value: " HorizontalAlignment="Right" VerticalAlignment="Center" Margin="3"/>
    <ComboBox Name="MaxRedComboBox" Grid.Row="1" Grid.Column="1" Foreground="Red" Width="50" HorizontalAlignment="Left" VerticalAlignment="Center" HorizontalContentAlignment="Right"/>
    <ComboBox Name="MaxGreenComboBox" Grid.Row="1" Grid.Column="2" Foreground="Green" Width="50" HorizontalAlignment="Left" VerticalAlignment="Center" HorizontalContentAlignment="Right"/>
    <ComboBox Name="MaxBlueComboBox" Grid.Row="1" Grid.Column=
```

```xml
"3" Foreground= "Blue" Width= "50" HorizontalAlignment= "Left" VerticalAlignment="Center" HorizontalContentAlignment="Right"/>
    </Grid>
     <Grid Name= "PercentClipParamsGrid" Visibility= "Visible" Margin="3">
        <Grid.RowDefinitions>
            <RowDefinition/>
            <RowDefinition/>
        </Grid.RowDefinitions>
        <Grid.ColumnDefinitions>
            <ColumnDefinition/>
            <ColumnDefinition Width="3*"/>
        </Grid.ColumnDefinitions>
        <TextBlock Grid.Row="0" Grid.Column="0" Text="Minimum %: " HorizontalAlignment="Center" VerticalAlignment="Center"/>
        <Slider Name="MinimumValueSlider" Grid.Row="0" Grid.Column="1" VerticalAlignment="Center" Minimum="0" Maximum="100">
            <Slider.ToolTip>
                <ToolTip Content="{Binding RelativeSource={RelativeSource Self},Path=PlacementTarget.Value}"/>
            </Slider.ToolTip>
        </Slider>
        <TextBlock Grid.Row="1" Grid.Column="0" Text="Maximum %: " HorizontalAlignment="Center" VerticalAlignment="Center"/>
         <Slider Name="MaximumValueSlider" Grid.Row="1" Grid.Column="1" VerticalAlignment="Center" Minimum="0" Maximum="100">
            <Slider.ToolTip>
                 <ToolTip Content=" {Binding RelativeSource={RelativeSource Self},Path=PlacementTarget.Value}"></ToolTip>
            </Slider.ToolTip>
        </Slider>
    </Grid>
    <Grid Name="StdDevParamsGrid" Visibility="Visible">
        <Grid.RowDefinitions>
            <RowDefinition/>
        </Grid.RowDefinitions>
        <Grid.ColumnDefinitions>
            <ColumnDefinition/>
```

```xml
            <ColumnDefinition Width="3*"/>
        </Grid.ColumnDefinitions>
        <TextBlock Grid.Row="0" Grid.Column="0" Text="Factor: " HorizontalAlignment="Right" VerticalAlignment="Center"/>
        <ComboBox Name="StdDeviationFactorComboBox" Grid.Row="0" Grid.Column="1" MinWidth="60" VerticalAlignment="Center" HorizontalAlignment="Left"/>
    </Grid>
    <Button Name="ApplyRgbRendererButton" Margin="3" Content="Apply renderer" Click="ApplyRgbRendererButton_Click"/>
</StackPanel>
```

4. 修改 MainWindow.xaml.cs 代码

添加命名空间:

```csharp
using Esri.ArcGISRuntime.Mapping;
using Esri.ArcGISRuntime.Rasters;
```

添加全部变量, 存储栅格路径和栅格图层:

```csharp
string _rasterPath = @"raster\北部湾大学.tif";
RasterLayer _rasterLayer;
```

编写 InitMap 函数, 打开栅格数据, 设置视点, 添加到地图中显示:

```csharp
async void InitMap()
{
  string rasterFile = System.IO.Path.Combine(AppDomain.CurrentDomain.BaseDirectory, _rasterPath);
    //Load the raster file.
    Raster raster = new Raster(rasterFile);
    //Create a new raster layer to show the image.
    _rasterLayer = new RasterLayer(raster);
     await _rasterLayer.LoadAsync();
    //Set the initial viewpoint with the raster's full extent.
    mapView1.Map.InitialViewpoint = new Viewpoint(_rasterLayer.FullExtent);
    //Add the layer to the map.
    mapView1.Map.OperationalLayers.Add(_rasterLayer);
}
```

对 UI 界面初始化:

```
void InitUI()
{
    //Add available stretch types to the combo box.
    StretchTypeComboBox.Items.Add("Min Max");
    StretchTypeComboBox.Items.Add("Percent Clip");
    StretchTypeComboBox.Items.Add("Standard Deviation");
    //Select "Min Max" as the stretch type.
    StretchTypeComboBox.SelectedIndex = 0;
    //Create a range of values from 0-255.
    IEnumerable<int> minMaxValues = Enumerable.Range(0, 256);
    MinRedComboBox.ItemsSource = minMaxValues;
    MinRedComboBox.SelectedValue = 0;
    MaxRedComboBox.ItemsSource = minMaxValues;
    MaxRedComboBox.SelectedValue = 255;
    MinGreenComboBox.ItemsSource = minMaxValues;
    MinGreenComboBox.SelectedValue = 0;
    MaxGreenComboBox.ItemsSource = minMaxValues;
    MaxGreenComboBox.SelectedValue = 255;
    MinBlueComboBox.ItemsSource = minMaxValues;
    MinBlueComboBox.SelectedValue = 0;
    MaxBlueComboBox.ItemsSource = minMaxValues;
    MaxBlueComboBox.SelectedValue = 255;
    //Fill the standard deviation factor combo box.
    IEnumerable<int> wholeStdDevs = Enumerable.Range(1, 10);
    StdDeviationFactorComboBox.ItemsSource = wholeStdDevs.Select(i => (double)i /2);
    StdDeviationFactorComboBox.SelectedValue = 2.0;
}
```

响应拉伸类型选择变化：

```
private void StretchTypeComboBox_SelectionChanged(object sender, SelectionChangedEventArgs e)
{
    //Hide all UI controls for the input parameters.
    MinMaxParamsGrid.Visibility = Visibility.Collapsed;
    PercentClipParamsGrid.Visibility = Visibility.Collapsed;
    StdDevParamsGrid.Visibility = Visibility.Collapsed;
```

```csharp
        //show the corresponding input controls.
        switch (StretchTypeComboBox.SelectedValue.ToString())
        {
            case "Min Max":
                MinMaxParamsGrid.Visibility = Visibility.Visible;
                break;
            case "Percent Clip":
                PercentClipParamsGrid.Visibility = Visibility.Visible;
                break;
            case "Standard Deviation":
                StdDevParamsGrid.Visibility = Visibility.Visible;
                break;
        }
    }
```

应用影像拉伸：

```csharp
private void ApplyRgbRendererButton_Click(object sender, RoutedEventArgs e)
{
    StretchParameters stretchParameters = null;
    //apply the corresponding input parameters to create the renderer.
    switch (StretchTypeComboBox.SelectedValue.ToString())
    {
        case "Min Max":
            //minimum and maximum values for red, green, and blue bands.
            double minRed = Convert.ToDouble(MinRedComboBox.SelectedValue);
            double minGreen = Convert.ToDouble(MinGreenComboBox.SelectedValue);
            double minBlue = Convert.ToDouble(MinBlueComboBox.SelectedValue);
            double maxRed = Convert.ToDouble(MaxRedComboBox.SelectedValue);
            double maxGreen = Convert.ToDouble(MaxGreenComboBox.SelectedValue);
            double maxBlue = Convert.ToDouble(MaxBlueComboBox.SelectedValue);
```

```
            //Create an array of the minimum and maximum values.
            double[] minValues = { minRed, minGreen, minBlue };
            double[] maxValues = { maxRed, maxGreen, maxBlue };
            //Create a new MinMaxStretchParameters with the values.
            stretchParameters = new MinMaxStretchParameters(minValues, maxValues);
            break;
        case "Percent Clip":
            //Get the percentile cutoff.
            double minimumPercent = MinimumValueSlider.Value;
            double maximumPercent = MaximumValueSlider.Value;
            //Create a new PercentClipStretchParameters with the inputs.
            stretchParameters = new PercentClipStretchParameters(minimumPercent, maximumPercent);
            break;
        case "Standard Deviation":
            //Read the standard deviation factor.
            double standardDeviationFactor = Convert.ToDouble(StdDeviationFactorComboBox.SelectedValue);
            //Create a new StandardDeviationStretchParameters.
            stretchParameters = new StandardDeviationStretchParameters(standardDeviationFactor);
            break;
    }
    //Create an array to specify the raster bands (red, green, blue).
    int[] bands = { 0, 1, 2 };
    // Create the RgbRenderer with the stretch parameters created above.
    RgbRenderer rasterRenderer = new RgbRenderer(stretchParameters, bands, null, true);
    _rasterLayer.Renderer = rasterRenderer;
}
```

最后在构造函数中调用地图初始化和界面初始化：

```
public MainWindow()
{
    InitializeComponent();
```

```
    InitMap();
    InitUI();
}
```

【实验结果】

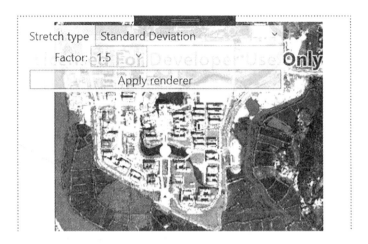

第4章　在线三维场景浏览

场景用于持久化和共享 ArcGIS 平台的 3D 内容。在 3D App 中可视化的数据在 Scene 类中定义。可以在场景中显示的数据类型包括以下三类。

（1）表面图层（surface layer），定义 3D 可视化时的高程信息。表面数据有多种形式，可以来自本地栅格数据例如 DEM 或 DTED，也可以来自 ArcGIS 影像服务。

（2）底图（basemap），可以覆盖在表面图层上，通过表面图层获取高程。

（3）业务图层（operational layer），包含矢量要素图层或场景图层。场景图层表达人工和自然 3D 内容，例如建筑、树木、山谷、山脉，可以来自场景服务或者本地场景图层包（scene layer package, slpk）。

第1节　高程服务场景三维浏览器

场景视图（SceneView）是三维地图控件，可以对高程和三维实体（建筑/树木等）进行三维展示。

【实验目的】

将在线高程服务加载到场景中进行三维可视化。

【实验数据】

Arcgis 在线高程服务：

http://elevation3d.arcgis.com/arcgis/rest/services/WorldElevation3D/Terrain3D/ImageServer

【实验步骤】

1. 新建项目

.NET 框架：与 ArcGIS Runtime SDK 对应
模板列表：Visual C#/Windows/Windows Classic Desktop
模板：ArcGIS Runtime Application（WPF）
名称：ElevationServiceSceneViewer
位置：D：\ArcGISRuntimeTutorial

2. 编译项目

详情见第 1 章第 3 节。

3. 修改 MainWindow.xaml 标记语言

去掉默认的地图视图,加入场景视图,添加名称属性:

```
<esri:SceneView Name="sceneView1">
    <esri:Scene />
</esri:SceneView>
```

设计 UI,添加工具条和子控件,其中包括两个按钮:

```
<ToolBar Margin="10" Opacity="0.8" HorizontalAlignment="Left" VerticalAlignment="Top">
    <Button Margin="3" Name="btnAdd" Content="Add" Click="btnAdd_Click" />
</ToolBar>
```

4. 修改 MainWindow.xaml.cs 代码

添加命名空间:

```
using Esri.ArcGISRuntime.Mapping;
```

定义全局变量,存储高程影像服务 url:

```
string _elevationImageService = "http://elevation3d.arcgis.com/arcgis/rest/services/WorldElevation3D/Terrain3D/ImageServer";
```

编写 InitMap 函数,添加底图:

```
void InitBasemap()
{
    //Add the imagery basemap to the scene's base map.
    sceneView1.Scene.Basemap = Basemap.CreateImageryWithLabels();
}
```

编写 InitScene,对场景进行初始化,其逻辑为:Uri→ArcGISTiledElevationSource→Surface→Scene→Camera

```
private void InitScene()
{
    Uri uri = new Uri(_elevationImageService);
    //Create an ArcGIS tiled elevation.
    ArcGISTiledElevationSource agtes = new ArcGISTiledElevationSource(uri);
    //Create a new surface.
    Surface surface = new Surface();
    //Add the ArcGIS tiled elevation source to the surface's elevated sources collection.
    surface.ElevationSources.Add(agtes);
    //Set the scene's base surface to the surface with the ArcGIS tiled
```

elevation source.
```
    sceneView1.Scene.BaseSurface = surface;
    //Create camera with an initial camera position (Mount Everest in the Alps mountains).
    Camera myCamera = new Camera(28.4, 83.9, 10010.0, 10.0, 80.0, 300.0);
    //Set the scene view's camera position.
    sceneView1.SetViewpointCameraAsync(myCamera);
}
```
响应添加场景按钮：
```
private void btnAdd_Click(object sender, RoutedEventArgs e)
{
    InitScene();
}
```
在构造函数中调用 Initialize 函数：
```
public MainWindow()
{
    InitializeComponent();
    InitMap ();
}
```

【实验结果】

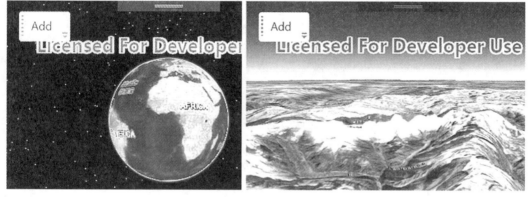

三维场景加载前后对比图

第 2 节　在线场景三维浏览器

场景视图 SceneView 是三维地图控件，可以对高程和三维实体(建筑/树木等)进行三维展示。

【实验目的】

将在线高程服务和建筑模型服务，加载到场景中进行三维可视化。

【实验数据】

ArcGIS 在线高程服务：

https：//scene.arcgis.com/arcgis/rest/services/BREST_DTM_1M/ImageServer

ArcGIS 在线建筑模型服务：

https：//scene.arcgis.com/arcgis/rest/services/Hosted/Buildings_Brest/SceneServer/0

【实验步骤】

1. 新建项目

.NET 框架：与 ArcGIS Runtime SDK 对应

模板列表：Visual C#/Windows/Windows Classic Desktop

模板：ArcGIS Runtime Application (WPF)

名称：ServiceSceneViewer

位置：D：\ArcGISRuntimeTutorial

2. 编译项目

详情见第 1 章第 3 节。

3. 修改 MainWindow.xaml 标记语言

去掉默认的地图视图，加入场景视图，添加名称属性：

```
<esri:SceneView Name="sceneView1">
    <esri:Scene />
</esri:SceneView>
```

设计 UI，添加工具条和子控件，其中包括两个按钮：

```
<ToolBar Margin="10" Opacity="0.8" HorizontalAlignment="Left" VerticalAlignment="Top">
    <Button Margin="3" Name="btnAdd" Content="Add" Click="btnAdd_Click"/>
</ToolBar>
```

4. 修改 MainWindow.xaml.cs 代码

添加命名空间：

```
using Esri.ArcGISRuntime.Mapping;
```

定义全局变量，存储高程影像服务 url 和建筑模型服务：

```
private Uri _elevationSourceUri = new Uri(@"https://scene.arcgis.com/arcgis/rest/services/BREST_DTM_1M/ImageServer");
//URL for the scene layer.
private Uri _buildUri = new Uri("https://scene.arcgis.com/arcgis/
```

rest/services/Hosted/Buildings_Brest/SceneServer/0");

在构造函数中加载底图：

```
public MainWindow()
{
    InitializeComponent();
    //Add the imagery basemap to the scene's base map.
    sceneView1.Scene.Basemap = Basemap.CreateImageryWithLabels();
}
```

编写 InitScene，对场景进行初始化：

```
private void InitScene()
{
    //Create and add an elevation source for the Scene.
    ArcGISTiledElevationSource elevationSrc=new ArcGISTiledElevationSource(_elevationSourceUri);
    sceneView1.Scene.BaseSurface.ElevationSources.Add(elevationSrc);
    //Create new scene layer from the url.
    ArcGISSceneLayer sceneLayer = new ArcGISSceneLayer(_buildUri);
    //Add created layer to the operational layers collection.
    sceneView1.Scene.OperationalLayers.Add(sceneLayer);
    //Create a camera with coordinates showing layer data.
    Camera camera = new Camera(48.378, -4.494, 200, 345, 65, 0);
    //Set view point of scene view using camera.
    sceneView1.SetViewpointCameraAsync(camera);
}
```

响应添加场景按钮：

```
private void btnAdd_Click(object sender, RoutedEventArgs e)
{
    InitScene();
}
```

【实验结果】

第 3 节　门户场景三维浏览器

在门户网站（Portal）发布的三维场景，通过分配 ID，可以统一入口，简化访问方式。

【实验目的】

开发 PortalSceneViewer，通过 Portal Item ID 浏览三维场景。

【实验数据】

Portal 在线三维场景服务：ID＝a13c3c3540144967bc933cb5e498b8e4

【实验步骤】

1. 新建项目

.NET 框架：与 ArcGIS Runtime SDK 对应
模板列表：Visual C#/Windows/Windows Classic Desktop
模板：ArcGIS Runtime Application（WPF）
名称：PortalSceneViewer
位置：D：\ArcGISRuntimeTutorial

2. 编译项目

详情见第 1 章第 3 节。

3. 修改 MainWindow.xaml 标记语言

添加名称属性：

```
<esri:SceneView Name="sceneView1">
    <esri:Scene />
</esri:SceneView>
```

设计 UI，添加堆叠容器，子控件包括：文本框和按钮。
文本框：输入 Portal Item ID。
按钮：加载 Portal Item 三维场景。

```
a13c3c3540144967bc933cb5e498b8e4
LoadPortal
```

修改 xaml 语言：

```xml
<StackPanel Background="Wheat" Opacity="0.9" Margin="3" VerticalAlignment="Top">
    <TextBox Name="tbPortalItem" Margin="3" Padding="3" MinWidth="200"/>
    <Button Name="btLoadPortal" Content="LoadPortal" Padding="3"
```

```
Margin="3" Click="btLoadPortal_Click"/>
</StackPanel>
```

4. 修改 MainWindow.xaml.cs 代码

添加命名空间：
```
using Esri.ArcGISRuntime.Mapping;
using Esri.ArcGISRuntime.Portal;
```
编写 Initialize 函数：
```
void Initialize()
{
    string portalItem = "a13c3c3540144967bc933cb5e498b8e4";
    tbPortalItem.Text = portalItem;
}
```
在构造函数中调用 Initialize 函数：
```
public MainWindow()
{
    InitializeComponent();
    Initialize();
}
```
响应按钮点击事件：
```
async void btLoadPortal_Click(object sender, RoutedEventArgs e)
{
    ArcGISPortal portal = await ArcGISPortal.CreateAsync();
    PortalItem item = await PortalItem.CreateAsync(portal, tbPortalItem.Text.Trim());
    Scene scene = new Scene(item);
    sceneView1.Scene = scene;
}
```

【实验结果】

第 4 节　Url 场景浏览器

在 WWW 上，每一个信息资源都有统一的地址，叫 URL(Uniform Resource Locator，统一资源定位器)。

URL 一般由 4 部分组成：协议、主机、端口、路径。一般语法格式为：

protocol：// hostname[：port] / path / [；parameters][？query]#fragment

【实验目的】

开发 UrlSceneViewer，通过 URL，直接创建三维场景。

【实验数据】

ArcGIS 在线三维场景服务：

string url = "https://www.arcgis.com/home/webscene/viewer.html?webscene=a13c3c3540144967bc933cb5e498b8e4";

string url2 = "https://www.arcgis.com/home/item.html? id=a13c3c3540144967bc933cb5e498b8e4";

string url3 = "https://www.arcgis.com/sharing/rest/content/items/a13c3c3540144967bc933cb5e498b8e4/data";

【实验步骤】

1. 新建项目

.NET 框架：与 ArcGIS Runtime SDK 对应

模板列表：Visual C#/Windows/Windows Classic Desktop

模板：ArcGIS Runtime Application（WPF）

名称：UrlSceneViewer

位置：D：\ArcGISRuntimeTutorial

2. 编译项目

详情见第 1 章第 3 节。

3. 修改 MainWindow.xaml 标记语言

添加名称属性：

```
<esri:SceneView Name="sceneView1">
    <esri:Scene />
</esri:SceneView>
```

设计 UI，添加堆叠容器，子控件包括：文本框和按钮。

文本框：输入 URL。

按钮：加载 URL 三维场景。

修改 xaml 语言：

```
<StackPanel Background="Wheat" Opacity="0.9" Margin="3"
VerticalAlignment="Top">
    <TextBox Name="tbUrl" Margin="3" Padding="3" MinWidth="200"/>
    <Button Name="btLoadUrl" Content="LoadUrl" Padding="3" Margin=
"3" Click="btLoadUrl_Click"/>
</StackPanel>
```

4. 修改 MainWindow.xaml.cs 代码

添加命名空间：

```
using Esri.ArcGISRuntime.Mapping;
```

编写 Initialize 函数：

```
void Initialize()
{
    string url="https://www.arcgis.com/home/webscene/viewer.html?webscene=a13c3c3540144967bc933cb5e498b8e4";
    string url2="https://www.arcgis.com/home/item.html?id=a13c3c3540144967bc933cb5e498b8e4";
    string url3=" https://www.arcgis.com/sharing/rest/content/items/a13c3c3540144967bc933cb5e498b8e4/data";
    tbUrl.Text = url;
}
```

在构造函数中调用 Initialize 函数：

```
public MainWindow()
{
    InitializeComponent();
    Initialize();
}
```

响应按钮点击事件：

```
void btLoadUrl_Click(object sender, RoutedEventArgs e)
{
    Scene scene = new Scene(new Uri(tbUrl.Text));
    sceneView1.Scene = scene;
}
```

【实验结果】

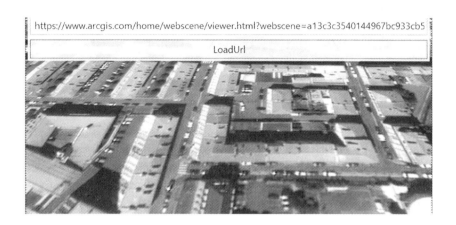

第5节　查看要素图层属性表

要素类一般都有属性表列出每个要素的属性信息,这是 GIS 与 CAD 的重要区别之一。

【实验目的】

(1)加载本地 SHP 文件,显示空间图形和对应的属性表;

(2)熟悉 DataGrid 控件的使用。

【实验数据】

ArcGISRuntimeSampleData \ 广西_shp \ 市界_region_Clip. shp

【实验步骤】

1. 新建项目

.NET 框架:与 ArcGIS Runtime SDK 对应

模板列表:Visual C#/Windows/Windows Classic Desktop

模板:ArcGIS Runtime Application(WPF)

名称:FeatureLayerAttributes

位置:D:\ArcGISRuntimeTutorial

2. 编译项目

详情见第1章第3节。

3. 修改 MainWindow. xaml 标记语言

对地图视图添加名称属性:

```
<esri:MapView Name = "mapView1" Map = "{Binding Map,Source ={Static
Resource MapViewModel}}" />
```

设计 UI：
```xml
<StackPanel Opacity="0.8">
    <TextBox Name="tbFile"/>
    <Button Name="tbLoadData" Content="LoadData" Click="tbLoadData_Click"/>
    <Button Name="btAttributeTable" Content="AttributeTable" Click="btAttributeTable_Click"/>
</StackPanel>
<DataGrid Name="dgFeaturesAttributeTable" VerticalAlignment="Bottom" Height="200" Visibility="Collapsed" Opacity="0.8"/>
```

4. 修改 MainWindow.xaml.cs 代码

添加命名空间：
```csharp
using Esri.ArcGISRuntime;
using Esri.ArcGISRuntime.Data;
using Esri.ArcGISRuntime.Mapping;
using System;
using System.Windows;
```
生成全局变量，用于存储 shp 文件路径：
```csharp
string _file = @"ArcGISRuntimeSampleData\广西_shp\市界_region_Clip.shp";
FeatureLayer _fl;
```
编写 Initialize 函数：
```csharp
void Initialize()
{
    string filepath = System.IO.Path.Combine(AppDomain.CurrentDomain.BaseDirectory + "..\\..\\..\\..\\", _file);
    tbFile.Text = filepath;
}
```
在构造函数中调用 Initialize 函数：
```csharp
public MainWindow()
{
    InitializeComponent();
    Initialize();
}
```
加载数据：
```csharp
async void tbLoadData_Click(object sender, RoutedEventArgs e)
{
```

第 5 节　查看要素图层属性表

```
    ShapefileFeatureTable sft = await ShapefileFeatureTable.OpenAsync
(tbFile.Text);
    //Create a feature layer to display the shapefile.
    _fl = new FeatureLayer(sft);
    await _fl.LoadAsync();
    if (_fl.LoadStatus == LoadStatus.Loaded)
      {
        //Add the feature layer to the map.
        mapView1.Map.OperationalLayers.Add(_fl);
      }
    //Set the map initial extent to the extent of the feature layer.
    //mapView1.Map.InitialViewpoint=new Viewpoint(_fl.FullExtent);
    await mapView1.SetViewpointGeometryAsync(_fl.FullExtent);
}
```

显示数据的属性表：

```
void btAttributeTable_Click(object sender, RoutedEventArgs e)
{
    dgFeaturesAttributeTable.ItemsSource = null;
    dgFeaturesAttributeTable.Columns.Clear();
    dgFeaturesAttributeTable.SetDataSourceAsync(_fl.FeatureTable);
    dgFeaturesAttributeTable.Visibility = Visibility.Visible;
}
```

【实验结果】

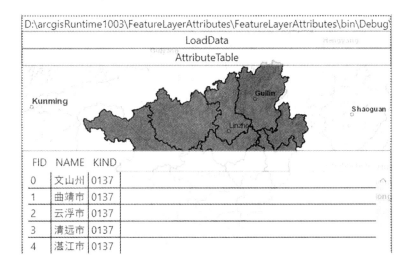

第6节 查看要素属性

属性与要素一一对应，可以实现"图文互查"。即通过点击图形，可以查询对应的属性；通过属性，可以定位图形。

【实验目的】

学习如何对属性进行查询。为了简化，本示例仅显示第一个要素的属性。

读者可以制作属性表导航工具栏，实现第 1 个(按钮)、上 1 个(按钮)、当前(文本框)、总数(标签)、下一个(按钮)、最后一个(按钮)要素的导航功能，设计 UI 如下：

|<　　<　　current　　total　　>　　>|

【实验数据】

本地 shp 文件：

ArcGISRuntimeSampleData \ 广西_shp \ 市界_region_Clip. shp

【实验步骤】

1. 新建项目

.NET 框架：与 ArcGIS Runtime SDK 对应

模板列表：Visual C#/Windows/Windows Classic Desktop

模板：ArcGIS Runtime Application（WPF）

名称：FeatureAttributes

位置：D:\ArcGISRuntimeTutorial

2. 编译项目

详情见第 1 章第 3 节。

3. 修改 MainWindow.xaml 标记语言

对地图视图，添加名称属性：

```
<esri:MapView Name="mapView1" Map="{Binding Map,Source={Static
Resource MapViewModel}}"/>
```

设计 UI：

```
<StackPanel Opacity="0.8">
    <TextBox Name="tbFile"/>
    <Button Name="tbLoadData" Content="LoadData" Click="tbLoadData
_Click"/>
    <Button Name="btAttributeTable" Content="AttributeTable"
Click="btAttributeTable_Click"/>
</StackPanel>
    <DataGrid Name="dgFeaturesAttributeTable" VerticalAlignment=
```

"Bottom" Height = "200" Visibility = "Collapsed" Opacity = "0.8"/>

4. 修改 MainWindow.xaml.cs 代码

添加命名空间:

```
using Esri.ArcGISRuntime;
using Esri.ArcGISRuntime.Data;
using Esri.ArcGISRuntime.Mapping;
using System;
using System.Windows;
```

生成全局变量,用于存储 shp 文件路径:

```
string _file = @ "ArcGISRuntimeSampleData\广西_shp\市界_region_Clip.shp";
FeatureLayer _fl;
```

编写 Initialize 函数:

```
void Initialize()
{
    string filepath = System.IO.Path.Combine(AppDomain.CurrentDomain.BaseDirectory + "..\\..\\..\\..\\", _file);
    tbFile.Text = filepath;
}
```

在构造函数中调用 Initialize 函数:

```
public MainWindow()
{
    InitializeComponent();
    Init();
}
```

加载数据:

```
async void tbLoadData_Click(object sender, RoutedEventArgs e)
{
    ShapefileFeatureTable sft = await ShapefileFeatureTable.OpenAsync(tbFile.Text);
    //Create a feature layer to display the shapefile.
    _fl = new FeatureLayer(sft);
    await _fl.LoadAsync();
    if (_fl.LoadStatus == LoadStatus.Loaded)
    {
        //Add the feature layer to the map.
        mapView1.Map.OperationalLayers.Add(_fl);
```

}
```
        //Set the map initial extent to the extent of the feature layer.
        //mapView1.Map.InitialViewpoint=new Viewpoint(_fl.FullExtent);
        await mapView1.SetViewpointGeometryAsync(_fl.FullExtent);
}
```

显示第一个要素的属性：

```
async void btAttributeTable_Click(object sender, RoutedEventArgs e)
{
    dgFeaturesAttributeTable.ItemsSource = null;
    dgFeaturesAttributeTable.Columns.Clear();
    Feature f = await _fl.FeatureTable.FirstOrDefaultAsync();
    if (f == null)
        return;
    dgFeaturesAttributeTable.SetDataSource(f);
    dgFeaturesAttributeTable.Visibility = Visibility.Visible;
    _fl.SelectFeature(f);
}
```

【实验结果】

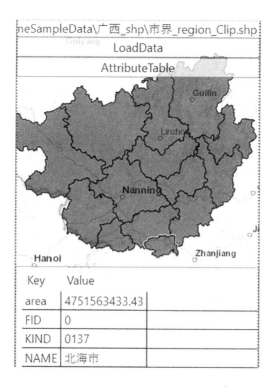

第 7 节　要素标注控制器

对要素进行标注，可以丰富地图的内容。不同的标注样式适合于不同的场景和不同的地图。ArcGIS Runtime 可以使用 JSON 语法进行标注定义。

【实验目的】

开发 FeatureLabelController，利用 JSON 字符串，实现标注样式的详细控制。

JSON 是字符串模型，不是面向对象的，容易出错。读者可以先将代码复制过去，运行成功后，再进行微调。

ArcGIS Runtime toolkits 中也提供了标注管理器的可视化控件，开发人员可以直接拿来用于快速开发，实现类似桌面 ArcMap、ArcGIS Pro 的效果。

【实验数据】

文件名：ChinaProvince \ bou2_4p. shp，中国省级行政区划。

字段：name，表示省(市)名称。

【实验步骤】

1. 新建项目

.NET 框架：与 ArcGIS Runtime SDK 对应

模板列表：Visual C#/Windows/Windows Classic Desktop

模板：ArcGIS Runtime Application（WPF）

名称：FeatureLabelController

位置：D:\ArcGISRuntimeTutorial

2. 编译项目

详情见第 1 章第 3 节。

3. 修改 MainWindow. xaml 标记语言

对地图视图添加名称属性：

`<esri:MapView Name="mapView1" Map="{Binding Map, Source={StaticResource MapViewModel}}"/>`

设计 UI，添加层叠面板，子控件包括文本框和按钮。

文本框，存储 shp 路径。

按钮，名称=btLabel，用于控制网格的生成、显示和隐藏。

```
<StackPanel>
```

```xml
<TextBox Name="tbFile" Opacity="0.8"/>
<Button Name="btLabel" Content="Label" Click="btLabel_Click" Opacity="0.8"/>
</StackPanel>
```

4. 修改 MainWindow.xaml.cs 代码

添加命名空间：

```csharp
using Esri.ArcGISRuntime.Data;
using Esri.ArcGISRuntime.Mapping;
using System.IO;
```

定义全局变量，存储 shp 文件位置：

```csharp
string _shpPath = @"ChinaProvince\bou2_4p.shp";
```

编写 Initialize 函数：

```csharp
private async void Initialize()
{
    //Get the path to the shapefile.
    string filepath = System.IO.Path.Combine(AppDomain.CurrentDomain.BaseDirectory, _shpPath);
    tbFile.Text = filepath;
    if (!File.Exists(filepath))
        return;
    //Open the shapefile.
    ShapefileFeatureTable myShapefile = await ShapefileFeatureTable.OpenAsync(filepath);
    //Create a feature layer to display the shapefile.
    FeatureLayer newFeatureLayer = new FeatureLayer(myShapefile);
    //Add the feature layer to the map.
    mapView1.Map.OperationalLayers.Add(newFeatureLayer);
    //Zoom the map to the extent of the shapefile.
    await mapView1.SetViewpointGeometryAsync(newFeatureLayer.FullExtent);
}
```

在构造函数中调用 Initialize 函数：

```csharp
public MainWindow()
{
    InitializeComponent();
    Initialize();
}
```

点击标注 Label 按钮，逻辑为：
如果要素图层没有生成标注，则通过 JSON，生成标注；
如果要素图层有标注，则切换标注的显示或者隐藏状态。
响应事件如下：

```
void btLabel_Click(object sender, RoutedEventArgs e)
{
    FeatureLayer fl = (mapView1.Map.OperationalLayers[0] as FeatureLayer);
    if (fl.LabelDefinitions.Count == 0)
    {
        string labelJson =
        @"{
""labelExpressionInfo"":{""expression"":""$feature.NAME""},
""labelPlacement"":""esriServerPolygonPlacementAlwaysHorizontal"",
        ""symbol"":
        {
            ""angle"":0,
            ""backgroundColor"":[0,0,0,0],
            ""borderLineColor"":[0,0,0,0],
            ""borderLineSize"":0,
            ""color"":[255,0,0,255],
            ""font"":
            {
                ""decoration"":""none"",
                ""size"":10,
                ""style"":""normal"",
                ""weight"":""normal""
            },
            ""haloColor"":[255,255,255,255],
            ""haloSize"":2,
            ""horizontalAlignment"":""center"",
            ""kerning"":false,
            ""type"":""esriTS"",
            ""verticalAlignment"":""middle"",
            ""xoffset"":0,
            ""yoffset"":0
        }
    }";
```

```
            LabelDefinition ld = LabelDefinition.FromJson(labelJson);
            fl.LabelDefinitions.Add(ld);
            fl.LabelsEnabled = true;
        }
        else
            fl.LabelsEnabled = ! fl.LabelsEnabled;
}
```

第5章 查询和编辑

要素类包括图形和属性,查询和编辑是要素类的基本功能。对属性进行查询和更新,对图形进行空间查询和编辑,是专业 GIS App 的基本功能。

对图形进行查询,一般通过 SQL where 字句进行筛选;对属性进行更新,表现为对字段值的编辑。

对图形进行查询,需要先给出坐标,一般通过鼠标绘制点、线、面进行选择;对图形进行编辑,简单的操作是直接绘制新的图形;复杂的操作,表现为对节点的交互动态编辑。

ArcGIS 还提供符号 Graphic,这是一种临时的、动态的、可高度定制显示样式的空间实体表示方法,常用来表示运动的车辆、瞬时风向、鼠标点击的位置等。Graphic 由开发人员手动进行管理,如果需要自动持久化,则需转化为要素类。

第1节 要素图层查询器

要素图层显示来自要素服务或者本地数据源,例如 shp 文件、地理包(GeoPackage)和地理数据库(Geodatabase)。要素图层能被用于显示、查找、查询。

【实验目的】

用户在文本框中输入关键词,点击查询按钮,在要素表中执行查询,将结果添加进选择集中并高亮显示,同时将地图缩放至当前选择要素。

【实验数据】

ArcGIS Server sample web service 在线地图,美国各州矢量在线地图:
http://sampleserver6.arcgisonline.com/arcgis/rest/services/USA/MapServer/2

【实验步骤】

1. 新建项目

.NET 框架:与 ArcGIS Runtime SDK 对应
模板列表:Visual C#/Windows/Windows Classic Desktop
模板:ArcGIS Runtime Application(WPF)
名称:FeatureLayerQueryer
位置:D:\ArcGISRuntimeTutorial

2. 编译项目

详情见第 1 章第 3 节。

3. 修改 MainWindow.xaml 标记语言

将网格分为两行，上面一行为工具条，工具条包括一个查询文本框和查询按钮；下面一行为地图视图区。

```
<Grid>
    <Grid.RowDefinitions>
        <RowDefinition Height="auto"/>
        <RowDefinition Height="*"/>
    </Grid.RowDefinitions>
    <ToolBar Grid.Row="0">
        <TextBox Name="tbQuery" Text="New York" Width="200"/>
        <Button Content="query" Name="btnQuery" Click="btnQuery_Click"/>
    </ToolBar>
    <esri:MapView Grid.Row="1" Name="mapView1" Map="{Binding Map, Source={StaticResource MapViewModel}}"/>
</Grid>
```

4. 修改 MainWindow.xaml.cs 代码

添加命名空间：

```
using Esri.ArcGISRuntime.Data;
using Esri.ArcGISRuntime.Geometry;
using Esri.ArcGISRuntime.Mapping;
using Esri.ArcGISRuntime.Symbology;
```

定义两个全局变量：

```
//Create reference to service of US States.
string _statesUrl = "http://sampleserver6.arcgisonline.com/arcgis/rest/services/USA/MapServer/2";
//Create globally available feature layer for easy referencing.
FeatureLayer _featureLayer;
```

编写 Init 函数：

```
void Init()
{
    //通过 url 生成要素表。
    Uri uri = new Uri(_statesUrl);
```

```
    ServiceFeatureTable featureTable = new ServiceFeatureTable(uri);
    //使用要素表生成要素图层。
    _featureLayer = new FeatureLayer(featureTable);
    _featureLayer.Opacity = 0.6;
    //构建图层渲染符号。
    SimpleLineSymbol lineSymbol = new SimpleLineSymbol(SimpleLineSymbolStyle.Solid, Colors.Black, 1);
    SimpleFillSymbol fillSymbol = newSimpleFillSymbol(SimpleFillSymbolStyle.Solid, Colors.Yellow, lineSymbol);
    //设置图层渲染符号。
    _featureLayer.Renderer = new SimpleRenderer(fillSymbol);
    _featureLayer.Loaded += (s, e) ⇒ Dispatcher.Invoke(() ⇒ mapView1.SetViewpointAsync(new Viewpoint(_featureLayer.FullExtent)));
    mapView1.Map.OperationalLayers.Add(_featureLayer);
}
```

编写查询函数,这是本实验的重点:

```
private async Task QueryStateFeature(string stateName)
{
    //生成查询参数用于要素表查询。
    QueryParameters queryParams = new QueryParameters();
    //去掉前后空格并转换为大写。
    String formattedStateName = stateName.Trim().ToUpper();
    //构建查询表达式。
    queryParams.WhereClause = "upper(STATE_NAME) LIKE '% " + formattedStateName + "%'";
    //查询要素表。
    FeatureQueryResult queryResult = await _featureLayer.FeatureTable.QueryFeaturesAsync(queryParams);
    //查询结果转换为列表。
    var features = queryResult.ToList();
    if (features.Any())
    {
        //获取查询结果中第一个要素。
        Feature feature = features[0];
        //将返回要素添加到当前选择集中。
        _featureLayer.SelectFeature(feature);
        //缩放到当前选择要素。
```

```
            await mapView1.SetViewpointGeometryAsync(feature.Geometry.
Extent);
        }
}
    响应用户点击事件：
private async void btnQuery_Click(object sender, RoutedEventArgs e)
{
    //清空以前的选择集合。
    _featureLayer.ClearSelection();
    //执行查询过程。
    await QueryStateFeature(tbQuery.Text);
}
    最后在构造函数中调用 Init 函数：
public MainWindow()
{
    InitializeComponent();
    Init();
}
```

【实验结果】

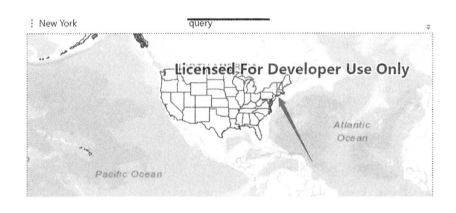

第 2 节　Shp 文件查询器

通过关键词，构建 SQL 语句，实现精确模型或模型查询，实现"以文查图"。

【实验目的】

在中国省级地图中，查询名称中包含"湖"的省份。湖南省和湖北省满足查询要求，在地图上高亮显示出来。

【实验数据】

本地 shp 文件，要素内容为小比例尺的中国各省级行政区划边界：
ArcGISRuntimeSampleData \ ChinaProvince \ bou2_4p. shp

【实验步骤】

1. 新建项目

.NET 框架：与 ArcGIS Runtime SDK 对应
模板列表：Visual C#/Windows/Windows Classic Desktop
模板：ArcGIS Runtime Application（WPF）
名称：ShapefileQueryer
位置：D：\ArcGISRuntimeTutorial

2. 编译项目

详情见第 1 章第 3 节。

3. 修改 MainWindow. xaml 标记语言

对地图视图，添加名称属性：
```
<esri:MapView Name="mapView1" Map="{Binding Map,Source={StaticResource MapViewModel}}" />
```
设计 UI：
```
<StackPanel>
<TextBox Name="tbShp" />
    <Button Content="LoadShp" Name="btLoadShp" Click="btLoadShp_Click" />
    <TextBox Name="tbQuery" />
    <Button Content="Query" Name="btQuery" Click="btQuery_Click" />
</StackPanel>
```

4. 修改 MainWindow. xaml. cs 代码

添加命名空间：
```
using Esri.ArcGISRuntime.Data;
using Esri.ArcGISRuntime.Mapping;
using System.IO;
```
生成全局变量，用于存储 shp 文件路径：
```
string _file = @"ArcGISRuntimeSampleData\ChinaProvince\bou2_4p.shp";
    string _where="湖";
```
编写 Init 函数：

```csharp
void Init()
{
    string filepath = System.IO.Path.Combine(AppDomain.CurrentDomain.
BaseDirectory + "..\\..\\..\\", _file);
    tbShp.Text = filepath;
    tbQuery.Text = _where;
}
```

在构造函数中调用 Init 函数:

```csharp
public MainWindow()
{
    InitializeComponent();
    Init();
}
```

加载 Shp 文件:

```csharp
async void btLoadShp_Click(object sender, RoutedEventArgs e)
{
    string filepath = tbShp.Text;
    if (!File.Exists(filepath))
        return;
    ShapefileFeatureTable sft = await
        ShapefileFeatureTable.OpenAsync(filepath);
    //Create a feature layer to display the shapefile.
    FeatureLayer fl = new FeatureLayer(sft);
    await fl.LoadAsync();
    mapView1.Map.OperationalLayers.Add(fl);
    mapView1.SetViewpoint(new Viewpoint(fl.FullExtent));
}
```

查询并选中:

```csharp
async void btQuery_Click(object sender, RoutedEventArgs e)
{
    //Create a query parameters that will be used to Query the feature table.
    QueryParameters queryParams = new QueryParameters();
    //Trim whitespace on the state name to prevent broken queries.
    String formattedStateName = tbQuery.Text.Trim().ToUpper();
    //Construct and assign the where clause that will be used to query the feature table.
    queryParams.WhereClause = $"name like '%{tbQuery.Text}%'";
```

```
    //Query the feature table.
    FeatureLayer fl = mapView1.Map.OperationalLayers[0] as Feature
Layer;
    //FeatureQueryResult queryResult = await
fl.FeatureTable.QueryFeaturesAsync(queryParams);
    await fl.SelectFeaturesAsync(queryParams, Esri.ArcGISRuntime.
Mapping.SelectionMode.New);
}
```

第3节 Sql 查询器

SQL 语句可以实现灵活的查询功能，适合熟悉数据库 SQL 语句的高级用户使用。

【实验目的】

由用户编写 SQL 查询语句，对本地文件进行查询和显示。

【实验数据】

本地 shp 文件，要素内容为小比例尺的中国各省级行政区划边界：ChinaProvince\bou2_4p.shp

【实验步骤】

1. 新建项目

.NET 框架：与 ArcGIS Runtime SDK 对应
模板列表：Visual C#/Windows/Windows Classic Desktop
模板：ArcGIS Runtime Application (WPF)
名称：SqlQuery
位置：D:\ArcGISRuntimeTutorial

2. 编译项目

详情见第1章第3节。

3. 修改 MainWindow.xaml 标记语言

对地图视图，添加名称属性：
`<esri:MapView Name="mapView1" Map="{Binding Map,Source={StaticResource MapViewModel}}"/>`

设计 UI：
```
<StackPanel Opacity="0.8">
    <TextBox Name="tbShp"/>
    <Button Content="LoadShp" Name="btLoadShp" Click="btLoadShp_Click"/>
```

```xml
<TextBox Name="tbQuery"/>
<Button Content="Query" Name="btQuery" Click="btQuery_Click"/>
</StackPanel>
```

4. 修改 MainWindow.xaml.cs 代码

添加命名空间：

```csharp
using Esri.ArcGISRuntime.Data;
using Esri.ArcGISRuntime.Mapping;
using System.IO;
```

生成全局变量，用于存储 shp 文件路径：

```csharp
string _file = @"ArcGISRuntimeSampleData\ChinaProvince\bou2_4p.shp";
string _where = $"name like '%广%'";
```

编写 Initialize 函数：

```csharp
void Initialize()
{
    string filepath = System.IO.Path.Combine(AppDomain.CurrentDomain.BaseDirectory + "..\\..\\..\\..\\", _file);
    tbShp.Text = filepath;
    tbQuery.Text = _where;
}
```

在构造函数中调用 Initialize 函数：

```csharp
public MainWindow()
{
    InitializeComponent();
    Initialize();
}
```

加载 Shp 文件：

```csharp
async void btLoadShp_Click(object sender, RoutedEventArgs e)
{
    string filepath = tbShp.Text;
    if (!File.Exists(filepath))
        return;
    ShapefileFeatureTable sft = await ShapefileFeatureTable.OpenAsync(filepath);
    //Create a feature layer to display the shapefile.
    FeatureLayer fl = new FeatureLayer(sft);
    await fl.LoadAsync();
```

```
    mapView1.Map.OperationalLayers.Add(fl);
    mapView1.SetViewpoint(new Viewpoint(fl.FullExtent));
}
```
SQL 查询：
```
async void btQuery_Click(object sender, RoutedEventArgs e)
{
    //Create a query parameters that will be used to Query the feature table.
    QueryParameters queryParams = new QueryParameters();
    //Trim whitespace on the province name to prevent broken queries.
    String formattedStateName = tbQuery.Text.Trim().ToUpper();
    //Construct and assign the where clause that will be used to query the feature table.
    queryParams.WhereClause = _where;
        FeatureLayer fl = mapView1.Map.OperationalLayers[0] as FeatureLayer;
        await fl.SelectFeaturesAsync(queryParams, Esri.ArcGISRuntime.Mapping.SelectionMode.New);
}
```
【实验结果】

当用户在查询文本框输入查询语句 name like '%广%' 后，点击查询，可以选中广西和广东。

第4节　识别图层要素

GIS 软件都带有识别功能，用户点击鼠标，对应位置的要素将会被激活。

【实验目的】

识别并选中鼠标点击出的要素。

【实验数据】

本地 shp 工作空间，包括小比例尺的三个要素类，分别为中国各省级行政区划边界，省级行政中心和全国一级河流：

ArcGISRuntimeSampleData \ ChinaProvince

【实验步骤】

1. 新建项目

.NET 框架：与 ArcGIS Runtime SDK 对应

模板列表：Visual C#/Windows/Windows Classic Desktop

模板：ArcGIS Runtime Application（WPF）

名称：IdentifyLayers
位置：D：\ArcGISRuntimeTutorial

2. 编译项目

详情见第 1 章第 3 节。

3. 修改 MainWindow.xaml 标记语言

对地图视图，添加名称属性：

```
<esri:MapView Name="mapView1" Map="{Binding Map,Source={StaticResource MapViewModel}}"/>
```

设计 UI：

```
<StackPanel Opacity="0.8">
    <TextBox Name="tbWorkspace"/>
    <Button Name="btOpen" Content="Open" Click="btOpen_Click"/>
    <ToggleButton Content="IdentifyLayers"
        Name="tbIdentifyLayers"></ToggleButton>
</StackPanel>
```

4. 修改 MainWindow.xaml.cs 代码

添加命名空间：

```
using Esri.ArcGISRuntime.Data;
using Esri.ArcGISRuntime.Geometry;
using Esri.ArcGISRuntime.Mapping;
using Esri.ArcGISRuntime.UI;
```

生成全局变量，用于存储 shp 文件路径：

```
string _path = @"ArcGISRuntimeSampleData\ChinaProvince";
```

编写 Initialize 函数：

```
void Initialize()
{
    string workspace =System.IO.Path.Combine(AppDomain.CurrentDomain.BaseDirectory + "..\\..\\..\\..\\", _path);
    tbWorkspace.Text = workspace;
}
```

在构造函数中调用 Initialize 函数：

```
public MainWindow()
{
    InitializeComponent();
```

```csharp
    Initialize();
}
```

打开 shp 工作空间：

```csharp
async void btOpen_Click(object sender, RoutedEventArgs e)
{
    string filepath = tbWorkspace.Text.Trim();
    string[] files = System.IO.Directory.GetFiles(filepath, "*.shp");
    List<Envelope> extents = new List<Envelope>();
    foreach (string file in files)
    {
        ShapefileFeatureTable sft = await ShapefileFeatureTable.OpenAsync(file);
        //Create a feature layer to display the shapefile.
        FeatureLayer fl = new FeatureLayer(sft);
        await fl.LoadAsync();
        mapView1.Map.OperationalLayers.Add(fl);
        //Add the extent to the list of extents.
        extents.Add(fl.FullExtent);
    }
    // Use the geometry engine to compute the full extent of the ENC Exchange Set.
    Envelope fullExtent = GeometryEngine.CombineExtents(extents);
    //Set the viewpoint.
    mapView1.SetViewpoint(new Viewpoint(fullExtent));
}
```

触摸（点击）地图，切换选择状态：

```csharp
async void mapView1_GeoViewTapped(object sender, Esri.ArcGISRuntime.UI.Controls.GeoViewInputEventArgs e)
{
    if (! tbIdentifyLayers.IsChecked.Value)
        return;
    mapView1.Map.OperationalLayers.ClearSelection();
    Point screenPoint = e.Position;
    double tolerance = 10;
    bool popupsOnly = false;
    IReadOnlyList<IdentifyLayerResult> identifyResults = await mapView1.IdentifyLayersAsync(screenPoint, tolerance, popupsOnly);
```

```
foreach (IdentifyLayerResult result in identifyResults)
{
    FeatureLayer fl = result.LayerContent as FeatureLayer;
    foreach (GeoElement ge in result.GeoElements)
    {
        Feature feature = ge as Feature;
        fl.SelectFeature(feature);
    }
}
```

第 5 节　识 别 符 号

符号（Graphics）是对空间实体的可视化表达，不过是临时的，不是持久化的，这是与要素的主要区别。Graphics 常用来表示运动符号，例如运动的车辆；也可用来表示用户编辑过程中的草图。

【实验目的】

对绘制的 Graphics 进行识别。首次点击可以选中该物体，再次点击取消选择。

【实验数据】

要素内容为小比例尺的中国各省级行政中心：
ArcGISRuntimeSampleData\ChinaProvince\res1_4m.shp

【实验步骤】

1. 新建项目

.NET 框架：与 ArcGIS Runtime SDK 对应

模板列表：Visual C#/Windows/Windows Classic Desktop

模板：ArcGIS Runtime Application（WPF）

名称：IdentifyGraphics

位置：D：\ArcGISRuntimeTutorial

2. 编译项目

详情见第 1 章第 3 节。

3. 修改 MainWindow.xaml 标记语言

对地图视图，添加名称属性：
```
<esri:MapView Name="mapView1" GeoViewTapped="mapView1_GeoViewTapped" Map="{Binding Map, Source={StaticResource MapViewModel}}" />
```
设计 UI：

```xml
<StackPanel Opacity="0.8">
    <TextBox Name="tbShp"/>
    <TextBox Name="tbPic"/>
    <Button Content="LoadShp" Name="btLoadShp" Click="btLoadShp_Click"/>
    <ToggleButton Content="IdentifyGraphics"
        Name="tbIdentifyGraphics"></ToggleButton>
</StackPanel>
```

4. 修改 MainWindow.xaml.cs 代码

添加命名空间：

```
using Esri.ArcGISRuntime.Data;
using Esri.ArcGISRuntime.Mapping;
using Esri.ArcGISRuntime.Symbology;
using Esri.ArcGISRuntime.UI;
```

生成全局变量，用于存储 shp 文件路径：

```
string _file = @"ArcGISRuntimeSampleData\ChinaProvince\res1_4m.shp";
string _pic = @"ArcGISRuntimeSampleData\images\plane32East.png";
```

编写 Initialize 函数：

```
void Initialize()
{
    string filepath = System.IO.Path.Combine(AppDomain.CurrentDomain.BaseDirectory + "..\\..\\..\\..\\", _file);
    string picpath = System.IO.Path.Combine(AppDomain.CurrentDomain.BaseDirectory + "..\\..\\..\\..\\", _pic);
    tbShp.Text = filepath;
    tbPic.Text = picpath;}
```

在构造函数中调用 Initialize 函数：

```
public MainWindow()
{
    InitializeComponent();
    Initialize();
}
```

添加 shp 文件，打开表格，生成符号（graphic）：

```
sync void btLoadShp_Click(object sender, RoutedEventArgs e)
```

```csharp
{
    string filepath = tbShp.Text;
    if (! File.Exists(filepath))
        return;
    string picpath = tbPic.Text;
    if (! File.Exists(picpath))
        return;
    ShapefileFeatureTable sft = await ShapefileFeatureTable.OpenAsync(filepath);
    mapView1.GraphicsOverlays.Add(new GraphicsOverlay());
    FileStream fs = new FileStream(picpath, FileMode.Open);
    PictureMarkerSymbol pms = await PictureMarkerSymbol.CreateAsync(fs);
    Graphic[] graphics = await GraphicCollectionEx.CreateFromFeatureTableAsync(sft, pms);
    mapView1.GraphicsOverlays[0].Graphics.AddRange(graphics);
    FeatureLayer fl = new FeatureLayer(sft);
    await fl.LoadAsync();
    mapView1.SetViewpoint(new Viewpoint(fl.FullExtent));
}
```

触摸(点击)地图，选中符号(飞机)，如果符号处于选中状态，则取消选中。

```csharp
async void mapView1_GeoViewTapped(object sender, Esri.ArcGIS.Runtime.UI.Controls.GeoViewInputEventArgs e)
{
    if (! tbIdentifyGraphics.IsChecked.Value)
        return;
    Point screenPoint = e.Position;
    double tolerance = 10;
    bool popupsOnly = false;
    var identifyResults = await mapView1.IdentifyGraphicsOverlaysAsync(screenPoint, tolerance, popupsOnly);
    foreach (IdentifyGraphicsOverlayResult igor in identifyResults)
    {
        foreach (Graphic g in igor.Graphics)
            g.IsSelected = ! g.IsSelected;
    }
}
```

【实验结果】

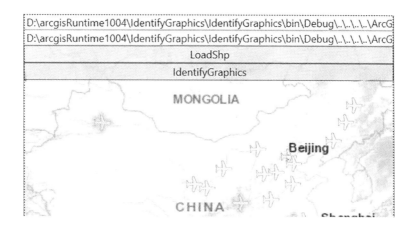

第6节 绘 制 点

在采集数据、对图形进行矢量化时,经常需要临时绘制符号。符号在客户端绘制,可以使用 PNG 透明图片表达,具有非常漂亮的样式。符号是临时性的,非常适合表达动态、需要实时更新的目标或现象。如果需要持久化,则需要添加到要素类中。本节介绍临时绘制符号中的一种绘制点。

【实验目的】

通过点击鼠标,绘制点符号。

【实验数据】

ArcGIS Online 在线底图服务:Runtime 默认底图服务。

【实验步骤】

1. 新建项目

.NET 框架:与 ArcGIS Runtime SDK 对应

模板列表:Visual C#/Windows/Windows Classic Desktop

模板:ArcGIS Runtime Application(WPF)

名称:PointSketch

位置:D:\ArcGISRuntimeTutorial

2. 编译项目

详情见第1章第3节。

3. 修改 MainWindow.xaml 标记语言

对地图视图,添加名称属性:

```xml
<esri:MapView Name="mapView1" Map="{Binding Map, Source={StaticResource MapViewModel}}" />
```

设计 UI：

```xml
<ToolBar VerticalAlignment="Top">
    <Button Content="DrawPoint" Name="btDrawPoint" Click="btDrawPoint_Click" />
    <Button Content="Reset" Name="btReset" Click="btReset_Click" />
</ToolBar>
```

4. 修改 MainWindow.xaml.cs 代码

添加命名空间：

```csharp
using Esri.ArcGISRuntime.Data;
using Esri.ArcGISRuntime.Mapping;
using Esri.ArcGISRuntime.Geometry;
using Esri.ArcGISRuntime.Symbology;
using Esri.ArcGISRuntime.UI;
```

编写 Initialize 函数：

```csharp
void Initialize()
{
    string file = @"ArcGISRuntimeSampleData\images\marker48.png";
    string url = System.IO.Path.Combine(AppDomain.CurrentDomain.BaseDirectory + "..\\..\\..\\..\\", file);
    PictureMarkerSymbol pms = new PictureMarkerSymbol(new Uri(url));
    GraphicsOverlay go = new GraphicsOverlay()
    { Renderer = new SimpleRenderer(pms) };
    mapView1.GraphicsOverlays.Add(go);
}
```

在构造函数中调用 Initialize 函数：

```csharp
public MainWindow()
{
    InitializeComponent();
    Initialize();
}
```

绘制点命令：

```csharp
async void btDrawPoint_Click(object sender, RoutedEventArgs e)
{
    //Let the user tap on the map view using the point sketch mode.
    SketchCreationMode scm = SketchCreationMode.Point;
```

```
    Geometry geometry = await mapView1.SketchEditor.StartAsync(scm,
false);
    //Create a graphic for the facility.
    Graphic graphic = new Graphic(geometry);
    mapView1.GraphicsOverlays[0].Graphics.Add(graphic);
}
```

清空已经绘制的点图形：

```
void btReset_Click(object sender, RoutedEventArgs e)
{
    mapView1.GraphicsOverlays[0].Graphics.Clear();
}
```

【实验结果】

第 7 节　绘　制　线

在采集数据、对图形进行矢量化的时候，经常需要临时绘制符号。符号在客户端绘制，可以使用 PNG 透明图片表达，具有非常漂亮的样式。符号是临时性的，非常适合表达动态，需要实时更新的目标或现象。如果需要持久化，则需要添加到要素类中。

【实验目的】

通过点击鼠标，绘制连续的线符号。

【实验数据】

ArcGIS Online 在线底图服务，Runtime 默认底图服务。

【实验步骤】

1. 新建项目

.NET 框架：与 ArcGIS Runtime SDK 对应

模板列表：Visual C#/Windows/Windows Classic Desktop

模板：ArcGIS Runtime Application（WPF）

名称：LineSketch

第 5 章　查询和编辑

位置：D：\ArcGISRuntimeTutorial

2. 编译项目

详情见第 1 章第 3 节。

3. 修改 MainWindow.xaml 标记语言

对地图视图，添加名称属性：

```
<esri:MapView Name="mapView1" Map="{Binding Map,Source={StaticResource MapViewModel}}"/>
```

设计 UI：

```
<ToolBar VerticalAlignment="Top">
    <Button Content="DrawLine" Name="btDrawLine" Click="btDrawLine_Click"/>
    <Button Content="Reset" Name="btReset" Click="btReset_Click"/>
</ToolBar>
```

4. 修改 MainWindow.xaml.cs 代码

添加命名空间：

```
using Esri.ArcGISRuntime.Geometry;
using Esri.ArcGISRuntime.Symbology;
using Esri.ArcGISRuntime.UI;
using System.Drawing;
```

生成全局变量，用于存储 shp 文件路径：

```
string _file = @"ChinaProvince\bou2_4p.shp";
```

编写 Initialize 函数：

```
void Initialize()
{
    SimpleLineSymbol sls = new SimpleLineSymbol(SimpleLineSymbolStyle.Solid, Color.Black, 5);
    GraphicsOverlay go = new GraphicsOverlay()
        {Renderer = new SimpleRenderer(sls)};
    mapView1.GraphicsOverlays.Add(go);
    mapView1.GeoViewDoubleTapped += (s, e) =>
    {
        //If the sketch editor complete command is enabled, a sketch is in progress.
        if (mapView1.SketchEditor.CompleteCommand.CanExecute(null))
```

```
        // Set the event as handled.
            e.Handled = true;
    };
}
```

在构造函数中调用 Initialize 函数：
```
public MainWindow()
{
    InitializeComponent();
    Initialize ();
}
```

绘制线命令：
```
async void btDrawLine_Click(object sender, RoutedEventArgs e)
{
    // Let the user draw on the map view using the polyline sketch mode.
    SketchCreationMode creationMode = SketchCreationMode.Polyline;
    Geometry geometry = await mapView1.SketchEditor.StartAsync(creationMode, false);
    // Create the graphic.
    Graphic graphic = new Graphic(geometry);
    // Add a graphic from the polyline the user drew.
    mapView1.GraphicsOverlays[0].Graphics.Add(graphic);
}
```

清空已经绘制的线条：
```
void btReset_Click(object sender, RoutedEventArgs e)
{
    mapView1.GraphicsOverlays[0].Graphics.Clear();
}
```

【实验结果】

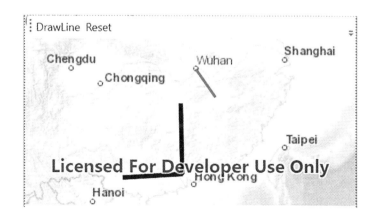

第 8 节　绘 制 多 边 形

在采集数据、对图形进行矢量化的时候，经常需要临时绘制符号。符号在客户端绘制，可以使用 PNG 透明图片表达，具有非常漂亮的样式。符号是临时性的，非常适合表达动态，需要实时更新的目标或现象。如果需要持久化，则需要添加到要素类中。

【实验目的】

通过点击鼠标，绘制封闭的多边形符号。

【实验数据】

ArcGIS Online 在线底图服务，Runtime 默认底图。

【实验步骤】

1. 新建项目

.NET 框架：与 ArcGIS Runtime SDK 对应

模板列表：Visual C#/Windows/Windows Classic Desktop

模板：ArcGIS Runtime Application（WPF）

名称：PolygonSketch

位置：D：\ArcGISRuntimeTutorial

2. 编译项目

详情见第 1 章第 3 节。

3. 修改 MainWindow.xaml 标记语言

对地图视图，添加名称属性：

```
<esri:MapView Name = "mapView1" Map = "{Binding Map, Source = {Static Resource MapViewModel}}" />
```

设计 UI：

```
<ToolBar VerticalAlignment = "Top">
    <Button Content = "DrawPolygon" Name = "btDrawPolygon" Click = "btDrawPolygon_Click" />
    <Button Content = "Reset" Name = "btReset" Click = "btReset_Click" />
</ToolBar>
```

4. 修改 MainWindow.xaml.cs 代码

添加命名空间：

```
using Esri.ArcGISRuntime.Geometry;
using Esri.ArcGISRuntime.Symbology;
using Esri.ArcGISRuntime.UI;
```

```
using System.Drawing;
```
编写 Initialize 函数：
```
void Initialize()
{
    SimpleLineSymbol sls = new SimpleLineSymbol(SimpleLineSymbolStyle.Solid, Color.Black, 2);
    SimpleFillSymbol sfs = new SimpleFillSymbol(SimpleFillSymbolStyle.ForwardDiagonal, Color.Black, sls);
    SimpleRenderer sr = new SimpleRenderer(sfs);
    GraphicsOverlay go = new GraphicsOverlay() { Renderer = sr };
    mapView1.GraphicsOverlays.Add(go);
    mapView1.GeoViewDoubleTapped += MapView1_GeoViewDoubleTapped;
}
```
在构造函数中调用 Initialize 函数：
```
public MainWindow()
{
    InitializeComponent();
    Initialize();
}
```
响应用户双击（鼠标或手势）：
```
void MapView1_GeoViewDoubleTapped(object sender, Esri.ArcGISRuntime.UI.Controls.GeoViewInputEventArgs e)
{
    if (mapView1.SketchEditor.CompleteCommand.CanExecute(null))
        e.Handled = true;
}
```
绘制多边形：
```
async void btDrawPolygon_Click(object sender, RoutedEventArgs e)
{
    Geometry geometry = await mapView1.SketchEditor.StartAsync(SketchCreationMode.Polygon, false);
    Graphic graphic = new Graphic(geometry);
    mapView1.GraphicsOverlays[0].Graphics.Add(graphic);
}
```
重置按钮，清除绘制内容：
```
void btReset_Click(object sender, RoutedEventArgs e)
{
    mapView1.GraphicsOverlays[0].Graphics.Clear();
}
```

【实验结果】

使用方法：点击 DrawLine，绘制多边形，双击结束。

点击 Reset，重置(清空)已经绘制的多边形。

第 9 节　点击选择要素

点击选择即点选。交互选择包括点选、画框选择、画线选择、多边形选择等。本节介绍点击选择要素。

【实验目的】

实现点选功能：添加本地 SHP 文件后，通过点击鼠标，选中所在位置的要素。

【实验数据】

ArcGIS Online 在线底图服务：

https：//sampleserver6.arcgisonline.com/arcgis/rest/services/World_Street_Map/MapServer

本地 shp 文件，要素内容为小比例尺的中国各省级行政区划边界：

ChinaProvince \ bou2_4p.shp

【实验步骤】

1. 新建项目

.NET 框架：与 ArcGIS Runtime SDK 对应

模板列表：Visual C#/Windows/Windows Classic Desktop

模板：ArcGIS Runtime Application（WPF）

名称：SelectByPoint，

位置：D：\ArcGISRuntimeTutorial

2. 编译项目

详情见第 1 章第 3 节。

3. 修改 MainWindow.xaml 标记语言

对地图视图,添加名称属性:

```
<esri:MapView GeoViewTapped="mapView1_GeoViewTapped" Name="mapView1" Map="{Binding Map, Source={StaticResource MapViewModel}}" />
```

设计 UI:

```
<StackPanel Opacity="0.8">
    <TextBox Name="tbShp" />
    <Button Content="LoadShp" Name="btLoadShp" Click="btLoadShp_Click" />
    <ToggleButton Content="SelectByPoint" Name="btSelectByPoint" Click="btSelectByPoint_Click" />
</StackPanel>
```

4. 修改 MainWindow.xaml.cs 代码

添加命名空间:

```
using Esri.ArcGISRuntime.Data;
using Esri.ArcGISRuntime.Mapping;
using System.IO;
```

生成全局变量,用于存储 shp 文件路径:

```
string _file = @"ArcGISRuntimeSampleData\ChinaProvince\bou2_4p.shp";
```

编写 Initialize 函数:

```
void Initialize()
{
    string filepath = System.IO.Path.Combine(AppDomain.CurrentDomain.BaseDirectory + "..\\..\\..\\..\\", _file);
    tbShp.Text = filepath;
}
```

在构造函数中调用 Initialize 函数:

```
public MainWindow()
{
    InitializeComponent();
    Initialize();
}
```

加载本地 Shp 文件:

```
async void btLoadShp_Click(object sender, RoutedEventArgs e)
{
```

```csharp
    string filepath = tbShp.Text;
    if (! File.Exists(filepath))
        return;
    ShapefileFeatureTable sft = await ShapefileFeatureTable.OpenAsync(filepath);
    //Create a feature layer to display the shapefile.
    FeatureLayer fl = new FeatureLayer(sft);
    await fl.LoadAsync();
    mapView1.Map.OperationalLayers.Add(fl);
    mapView1.SetViewpoint(new Viewpoint(fl.FullExtent));
}
```

点击地图选择要素：

```csharp
async void mapView1_GeoViewTapped(object sender, Esri.ArcGISRuntime.UI.Controls.GeoViewInputEventArgs e)
{
    if (! btSelectByPoint.IsChecked.Value)
        return;
    //Create a query parameters that will be used to Query the feature table.
    QueryParameters queryParams = new QueryParameters();
    // Geometry geometry = await mapView1.SketchEditor.StartAsync(SketchCreationMode.Point, false);
    queryParams.Geometry = e.Location;
    queryParams.SpatialRelationship = SpatialRelationship.Intersects;
    //Query the feature table.
    FeatureLayer fl = mapView1.Map.OperationalLayers[0] as FeatureLayer;
    //FeatureQueryResult queryResult = await fl.FeatureTable.QueryFeaturesAsync(queryParams);
    await fl.SelectFeaturesAsync(queryParams, Esri.ArcGISRuntime.Mapping.SelectionMode.New);
}
```

第 10 节　画线选择要素

画线选择也是交互选择中的一种。交互选择包括点选、拉框选择、画线选择，多边形选择等。

第 10 节 画线选择要素

【实验目的】

实现线选：添加本地 SHP 文件后，通过绘制 polyline，选中所在位置的要素。

【实验数据】

ArcGIS Online 在线底图服务：

https：//sampleserver6.arcgisonline.com/arcgis/rest/services/World_Street_Map/MapServer

本地 shp 文件，要素内容为小比例尺的中国各省级行政区划边界：

ChinaProvince \ bou2_4p.shp

【实验步骤】

1. 新建项目

.NET 框架：与 ArcGIS Runtime SDK 对应

模板列表：Visual C#/Windows/Windows Classic Desktop

模板：ArcGIS Runtime Application（WPF）

名称：SelectByLine

位置：D：\ArcGISRuntimeTutorial

2. 编译项目

详情见第 1 章第 3 节。

3. 修改 MainWindow.xaml 标记语言

对地图视图，添加名称属性：

```
<esri:MapView Name="mapView1" Map="{Binding Map,Source={StaticResource MapViewModel}}"/>
```

设计 UI：

```
<StackPanel Opacity="0.8">
    <TextBox Name="tbShp"/>
    <Button Content="Open" Name="btOpen" Click="btOpen_Click"/>
    <Button Content="Select" Name="btSelect" Click="btSelect_Click"/>
</StackPanel>
```

4. 修改 MainWindow.xaml.cs 代码

添加命名空间：

```
using Esri.ArcGISRuntime.Data;
using Esri.ArcGISRuntime.Geometry;
using Esri.ArcGISRuntime.Mapping;
using Esri.ArcGISRuntime.UI;
using System.IO;
```

注释掉命名空间：
```
//using System.Windows.Media;
```
生成全局变量，用于存储 shp 文件路径：
```
string _file = @"ArcGISRuntimeSampleData\ChinaProvince\bou2_4p.shp";
```
编写 Initialize 函数：
```
void Initialize()
{
    string filepath = System.IO.Path.Combine(AppDomain.CurrentDomain.BaseDirectory + "..\\..\\..\\..\\", _file);
    tbShp.Text = filepath;
}
```
在构造函数中调用 Initialize 函数：
```
public MainWindow()
{
    InitializeComponent();
    Initialize();
}
```
加载本地 Shp 文件：
```
async void btOpen_Click(object sender, RoutedEventArgs e)
{
    string filepath = tbShp.Text;
    if (!File.Exists(filepath))
        return;
    ShapefileFeatureTable sft = await ShapefileFeatureTable.OpenAsync(filepath);
    //Create a feature layer to display the shapefile.
    FeatureLayer fl = new FeatureLayer(sft);
    await fl.LoadAsync();
    mapView1.Map.OperationalLayers.Add(fl);
    mapView1.SetViewpoint(new Viewpoint(fl.FullExtent));
}
```
画线选择要素：
```
async void btSelect_Click(object sender, RoutedEventArgs e)
{
    QueryParameters queryParams = new QueryParameters();
    Geometry geometry = await mapView1.SketchEditor.StartAsync(SketchCreationMode.Polyline, false);
    queryParams.Geometry = geometry;
```

```
    queryParams.SpatialRelationship = SpatialRelationship.
Intersects;
    //Query the feature table.
    FeatureLayer fl = mapView1.Map.OperationalLayers[0] as Feature
Layer;
    //FeatureQueryResult queryResult = await fl.FeatureTable.Query
FeaturesAsync(queryParams);
    await fl.SelectFeaturesAsync(queryParams, Esri.ArcGISRuntime.
Mapping.SelectionMode.New);
}
```

第 11 节　画面选择要素

画面选择也属于交互选择的范畴。本节具体介绍画面选择要素。

【实验目的】

实现面选功能：添加本地 SHP 文件后，绘制多边形，选中所在位置的要素。

【实验数据】

ArcGIS Online 在线底图服务：

https：//sampleserver6.arcgisonline.com/arcgis/rest/services/World_Street_Map/MapServer

本地 shp 文件，要素内容为小比例尺的中国各省级行政区划边界：

ChinaProvince \ bou2_4p.shp

【实验步骤】

1. 新建项目

.NET 框架：与 ArcGIS Runtime SDK 对应

模板列表：Visual C#/Windows/Windows Classic Desktop

模板：ArcGIS Runtime Application（WPF）

名称：SelectByPolygon

位置：D：\ArcGISRuntimeTutorial

2. 编译项目

详情见第 1 章第 3 节。

3. 修改 MainWindow.xaml 标记语言

对地图视图，添加名称属性：

```
<esri:MapView Name="mapView1" Map="{Binding Map,Source={Static
Resource MapViewModel}}"/>
```

设计 UI：
```xml
<StackPanel Opacity="0.8">
    <TextBox Name="tbShp"/>
    <Button Content="Open" Name="btOpen" Click="btOpen_Click"/>
    <Button Content="Select" Name="btSelect" Click="btSelect_Click"/>
</StackPanel>
```

4. 修改 MainWindow.xaml.cs 代码

添加命名空间：
```csharp
using Esri.ArcGISRuntime.Data;
using Esri.ArcGISRuntime.Geometry;
using Esri.ArcGISRuntime.Mapping;
using Esri.ArcGISRuntime.UI;
using System.IO;
```

注释掉命名空间：
```csharp
//using System.Windows.Media;
```

生成全局变量，用于存储 shp 文件路径：
```csharp
string _file = @"ArcGISRuntimeSampleData\ChinaProvince\bou2_4p.shp";
```

编写 Initialize 函数：
```csharp
void Initialize()
{
    string filepath = System.IO.Path.Combine(AppDomain.CurrentDomain.BaseDirectory + "..\\..\\..\\..\\", _file);
    tbShp.Text = filepath;
}
```

在构造函数中调用 Initialize 函数：
```csharp
public MainWindow()
{
    InitializeComponent();
    Initialize();
}
```

加载本地 Shp 文件：
```csharp
async void btOpen_Click(object sender, RoutedEventArgs e)
{
    tring filepath = tbShp.Text;
    if (!File.Exists(filepath))
```

```
            return;
    ShapefileFeatureTable sft = await ShapefileFeatureTable.OpenAsync
(filepath);
    //Create a feature layer to display the shapefile.
    FeatureLayer fl = new FeatureLayer(sft);
    await fl.LoadAsync();
    mapView1.Map.OperationalLayers.Add(fl);
    mapView1.SetViewpoint(new Viewpoint(fl.FullExtent));
}
```

画面选择要素：

```
async void btSelect_Click(object sender, RoutedEventArgs e)
{
    QueryParameters queryParams = new QueryParameters();
     Geometry  geometry  =  await  mapView1.SketchEditor.StartAsync
(SketchCreationMode.Polygon, false);
    queryParams.Geometry = geometry;
    queryParams.SpatialRelationship = SpatialRelationship.
Intersects;
    //Query the feature table.
    FeatureLayer fl = mapView1.Map.OperationalLayers[0] as Feature
Layer;
    //FeatureQueryResult queryResult = await fl.FeatureTable.Query
FeaturesAsync(queryParams);
    await fl.SelectFeaturesAsync(queryParams, Esri.ArcGISRuntime.
Mapping.SelectionMode.New);
}
```

【实验结果】

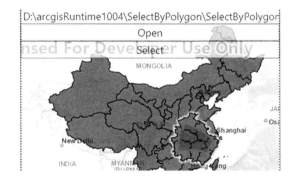

第 12 节 属 性 编 辑

GIS App 一般都提供属性的编辑功能。例如在采集空间信息的时候，提供属性的录入；在内业处理的时候，进行属性的更改。

【实验目的】

对选中的要素，在对话框中进行属性修改。

为了简化交互输入，直接使用了 VB. NET 中的文本输入对话框 Interaction. InputBox。

为了简化逻辑，系统模拟选择过程，自动选择第一个要素。开发人员可结合前面学习的识别功能、查询功能，将选择权交给用户。

【实验数据】

ArcGIS Online 在线底图服务：

https：//sampleserver6. arcgisonline. com/arcgis/rest/services/World_Street_Map/MapServer

本地 shp 文件，要素内容为小比例尺的中国各省级行政中心：ArcGISRuntime SampleData \ ChinaProvince \ res1_4m. shp

【实验步骤】

1. 新建项目

.NET 框架：与 ArcGIS Runtime SDK 对应

模板列表：Visual C#/Windows/Windows Classic Desktop

模板：ArcGIS Runtime Application（WPF）

名称：AttributeEditor

位置：D：\ArcGISRuntimeTutorial

2. 编译项目

详情见第 1 章第 3 节。

3. 修改 MainWindow. xaml 标记语言

对地图视图，添加名称属性：

```
<esri:MapView Name="mapView1" Map="{Binding Map,Source={StaticResource MapViewModel}}" />
```

设计 UI：

```
<StackPanel Opacity="0.8">
    <TextBox Name="tbFile" />
    <Button Name="btOpen" Content="Open" Click="btOpen_Click" />
```

```
    < Button  Content = " EditAttributes "  Name = " btEditAttributes "
Click="btEditAttributes_Click" />
</StackPanel>
```

4. 修改 MainWindow.xaml.cs 代码

添加命名空间：
```
using Esri.ArcGISRuntime.Data;
using Esri.ArcGISRuntime.Mapping;
using Esri.ArcGISRuntime.UI;
using Microsoft.VisualBasic;
using System.IO;
```

生成全局变量，用于存储 shp 文件路径：
```
string _file = @ " ArcGISRuntimeSampleData \ ChinaProvince \ res1 _
4m.shp";
```

编写 Initialize 函数：
```
void Initialize()
{
    string filepath = System.IO.Path.Combine(AppDomain.CurrentDomain.
BaseDirectory + "..\\..\\..\\..\\", _file);
    tbFile.Text = filepath;
}
```

在构造函数中调用 Initialize 函数：
```
public MainWindow()
{
    InitializeComponent();
    Initialize();
}
```

打开 shp 文件：
```
async void btOpen_Click(object sender, RoutedEventArgs e)
{
    string filepath = tbFile.Text;
    if (! File.Exists(filepath))
        return;
    ShapefileFeatureTable sft = await ShapefileFeatureTable.
```

```csharp
OpenAsync(filepath);
    //Create a feature layer to display the shapefile.
    FeatureLayer fl = new FeatureLayer(sft);
    await fl.LoadAsync();
    mapView1.Map.OperationalLayers.Add(fl);
    mapView1.SetViewpoint(new Viewpoint(fl.FullExtent));
    QueryParameters qp = new QueryParameters();
    qp.WhereClause="1=1";
    FeatureQueryResult fqr = await sft.QueryFeaturesAsync(qp);
    Feature feature = fqr.FirstOrDefault();
    if (feature == null)
        return;
    fl.SelectFeature(feature);
}
```

编辑属性：

```csharp
async void btEditAttributes_Click(object sender, RoutedEventArgs e)
{
    FeatureLayer fl = mapView1.Map.OperationalLayers[0] as FeatureLayer;
    FeatureQueryResult fqr = await fl.GetSelectedFeaturesAsync();
    Feature feature = fqr.FirstOrDefault();
    if (feature == null)
        return;
    var attributes = feature.Attributes;
    string defaultValue = feature.Attributes["PINYIN"].ToString();
    string newValue=Interaction.InputBox("new attribute", "attributes editor", defaultValue);
    if (string.IsNullOrEmpty(newValue) ||defaultValue == newValue)
        return;
    feature.Attributes["PINYIN"] = newValue;
    ShapefileFeatureTable sft = fl.FeatureTable as ShapefileFeatureTable;
    await sft.UpdateFeatureAsync(feature);
}
```

【实验结果】

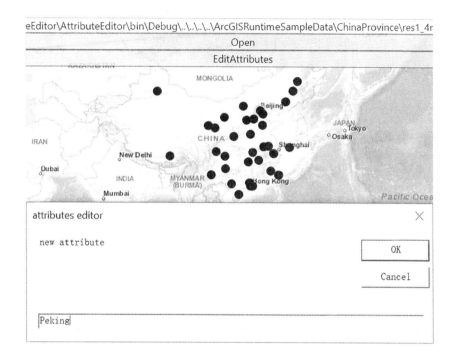

第13节 Shp点要素编辑

对空间要素进行编辑，是专业GIS的基本功能，区别于公共的电子地图系统。普通的公众电子图，例如百度地图、高德地图，只提供地图浏览，并不提供空间数据的编辑功能。

【实验目的】

点是最常见的要素几何类型，在公众地图中使用较多，常用来表示兴趣点（Point Of Interest）。通过点击，移动其几何位置。

【实验数据】

ArcGIS Online在线底图服务：

https：//sampleserver6.arcgisonline.com/arcgis/rest/services/World_Street_Map/MapServer

本地shp文件，要素内容为小比例尺的中国各省级行政区划边界：ChinaProvince\bou2_4p.shp

【实验步骤】

1. 新建项目

.NET框架：与ArcGIS Runtime SDK对应

模板列表：Visual C#/Windows/Windows Classic Desktop
模板：ArcGIS Runtime Application（WPF）
名称：ShpPointFeatureEditor
位置：D：\ArcGISRuntimeTutorial

2. 编译项目

详情见第 1 章第 3 节。

3. 修改 MainWindow.xaml 标记语言

对地图视图，添加名称属性：
```
<esri:MapView Name="mapView1" Map="{Binding Map,Source={StaticResource MapViewModel}}"/>
```
设计 UI：
```
<StackPanel Opacity="0.8">
    <TextBox Name="tbFile"/>
    <Button Name="btOpen" Content="Open" Click="btOpen_Click"/>
    <Button Content="Edit" Name="btEdit" Click="btEdit_Click"/>
</StackPanel>
```

4. 修改 MainWindow.xaml.cs 代码

添加命名空间：
```
using Esri.ArcGISRuntime.Data;
using Esri.ArcGISRuntime.Geometry;
using Esri.ArcGISRuntime.Mapping;
using Esri.ArcGISRuntime.UI;
using System.IO;
```
生成全局变量，用于存储 shp 文件路径：
```
string _file = @"ArcGISRuntimeSampleData\ChinaProvince\res1_4m.shp";
```
编写 Initialize 函数：
```
void Initialize()
{
    string filepath = System.IO.Path.Combine(AppDomain.CurrentDomain.BaseDirectory + "..\\..\\..\\..\\", _file);
    tbFile.Text = filepath;
}
```
在构造函数中调用 Initialize 函数：
```
public MainWindow()
```

```csharp
{
    InitializeComponent();
    Initialize ();
}
```

打开 shp 文件：
```csharp
async void btOpen_Click(object sender, RoutedEventArgs e)
{
    string filepath = tbFile.Text;
    if (! File.Exists(filepath))
        return;
    ShapefileFeatureTable sft = await ShapefileFeatureTable.OpenAsync(filepath);
    //Create a feature layer to display the shapefile.
    FeatureLayer fl = new FeatureLayer(sft);
    await fl.LoadAsync();
    mapView1.Map.OperationalLayers.Add(fl);
    mapView1.SetViewpoint(new Viewpoint(fl.FullExtent));
    QueryParameters qp = new QueryParameters();
    qp.WhereClause="1=1";
    FeatureQueryResult fqr = await sft.QueryFeaturesAsync(qp);
    Feature feature = fqr.FirstOrDefault();
    if (feature == null)
        return;
    fl.SelectFeature(feature);
}
```

点击地图，移动当前选中点：
```csharp
async void btEdit_Click(object sender, RoutedEventArgs e)
{
    Geometry geometry = await mapView1.SketchEditor.StartAsync(SketchCreationMode.Point, false);
    MapPoint mp = geometry as MapPoint;
    FeatureLayer fl = mapView1.Map.OperationalLayers[0] as FeatureLayer;
    FeatureQueryResult fqr = await fl.GetSelectedFeaturesAsync();
    Feature feature = fqr.FirstOrDefault();
    if (feature == null)
        return;
    feature.Geometry = geometry;
```

```
    ShapefileFeatureTable sft = fl.FeatureTable as ShapefileFeature
Table;
    await sft.UpdateFeatureAsync(feature);
}
```

【实验结果】

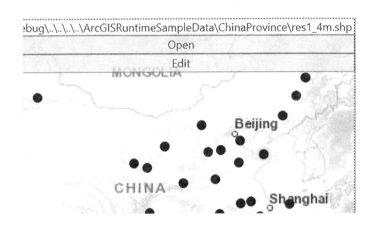

第 14 节　SHP 线要素编辑

线常用来表示忽略宽度的带状地图。例如中小地图上的道路、河流。注意在导航的大比例尺高精地图中，道路一般用多边形表示，以支撑未来的车道级导航。

【实验目的】

加载本地 SHP 线文件后，画线作为待编辑要素的几何图形。

【实验数据】

ArcGIS Online 在线底图服务：

https：//sampleserver6.arcgisonline.com/arcgis/rest/services/World_Street_Map/MapServer

本地 shp 文件，要素内容为小比例尺的水系：

ArcGISRuntimeSampleData \ ChinaProvince \ hyd1_4l.shp

【实验步骤】

1. 新建项目

.NET 框架：与 ArcGIS Runtime SDK 对应

模板列表：Visual C#/Windows/Windows Classic Desktop

模板：ArcGIS Runtime Application（WPF）

名称：ShpLineFeatureEditor

位置：D：\ArcGISRuntimeTutorial

2. 编译项目

详情见第 1 章第 3 节。

3. 修改 MainWindow.xaml 标记语言

对地图视图，添加名称属性：

```
<esri:MapView Name="mapView1" Map="{Binding Map,Source={StaticResource MapViewModel}}"/>
```

设计 UI：

```
<StackPanel Opacity="0.8">
    <TextBox Name="tbFile"/>
    <Button Name="btOpen" Content="Open" Click="btOpen_Click"/>
    <Button Content="Edit" Name="btEdit" Click="btEdit_Click"/>
</StackPanel>
```

4. 修改 MainWindow.xaml.cs 代码

添加命名空间：

```csharp
using Esri.ArcGISRuntime.Data;
using Esri.ArcGISRuntime.Geometry;
using Esri.ArcGISRuntime.Mapping;
using Esri.ArcGISRuntime.UI;
using System.IO;
//using System.Windows.Media;
```

生成全局变量，用于存储 shp 文件路径：

```csharp
string _file = @"ArcGISRuntimeSampleData\ChinaProvince\hyd1_4l.shp";
```

编写 Initialize 函数：

```csharp
void Initialize()
{
    string filepath = System.IO.Path.Combine(AppDomain.CurrentDomain.BaseDirectory + "..\\..\\..\\..\\", _file);
    tbFile.Text = filepath;
    mapView1.GeoViewDoubleTapped += (s, e) =>
    {
        if (mapView1.SketchEditor.CompleteCommand.CanExecute(null))
            e.Handled = true;
    };
}
```

在构造函数中调用 Initialize 函数：

```
public MainWindow()
{
    InitializeComponent();
    Initialize ();
}
```

打开地图：

```
async void btOpen_Click(object sender, RoutedEventArgs e)
{
    string filepath = tbFile.Text;
    if (! File.Exists(filepath))
        return;
    ShapefileFeatureTable sft = await ShapefileFeatureTable.OpenAsync(filepath);
    //Create a feature layer to display the shapefile.
    FeatureLayer fl = new FeatureLayer(sft);
    await fl.LoadAsync();
    mapView1.Map.OperationalLayers.Add(fl);
    mapView1.SetViewpoint(new Viewpoint(fl.FullExtent));
    QueryParameters qp = new QueryParameters();
    qp.WhereClause="1=1";
    FeatureQueryResult fqr = await sft.QueryFeaturesAsync(qp);
    Feature feature = fqr.FirstOrDefault();
    if (feature == null)
        return;
    fl.SelectFeature(feature);
}
```

编辑线：

```
async void btEdit_Click(object sender, RoutedEventArgs e)
{
    Geometry geometry = await mapView1.SketchEditor.StartAsync(SketchCreationMode.Polyline, false);
    MapPoint mp = geometry as MapPoint;
    FeatureLayer fl = mapView1.Map.OperationalLayers[0] as FeatureLayer;
    FeatureQueryResult fqr = await fl.GetSelectedFeaturesAsync();
    Feature feature = fqr.FirstOrDefault();
    if (feature == null)
```

```
            return;
        feature.Geometry = geometry;
        ShapefileFeatureTable sft = fl.FeatureTable as ShapefileFeature
Table;
        await sft.UpdateFeatureAsync(feature);
}
```

【实验结果】

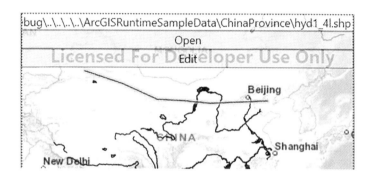

第 15 节　Shp 多边形要素编辑

多边形用于表达一定区域内的空间实体,具有周长和面积两个字段。在 GIS 系统中,周长和面积一般是系统字段,由数据库维护,属于只读。

【实验目的】

加载本地 SHP 多边形数据,系统模拟选中第一个要素;将用户绘制多边形,更新为待编辑要素的几何图形。

【实验数据】

ArcGIS Online 在线底图服务:

https://sampleserver6.arcgisonline.com/arcgis/rest/services/World_Street_Map/MapServer

本地 shp 文件,要素内容为小比例尺的水系:

ArcGISRuntimeSampleData \ ChinaProvince \ bou2_4p.shp

【实验步骤】

1. 新建项目

.NET 框架:与 ArcGIS Runtime SDK 对应

模板列表:Visual C#/Windows/Windows Classic Desktop

模板:ArcGIS Runtime Application (WPF)

名称：ShpPolygonFeatureEditor

位置：D：\ArcGISRuntimeTutorial

2. 编译项目

详情见第 1 章第 3 节。

3. 修改 MainWindow.xaml 标记语言

对地图视图，添加名称属性：

```
<esri:MapView Name=" mapView1 " Map=" {Binding Map, Source={StaticResource MapViewModel}}" />
```

设计 UI：

```
<StackPanel Opacity="0.8">
    <TextBox Name="tbFile" />
    <Button Name="btOpen" Content="Open" Click="btOpen_Click" />
    <Button Content="Edit" Name="btEdit" Click="btEdit_Click" />
</StackPanel>
```

4. 修改 MainWindow.xaml.cs 代码

添加命名空间：

```csharp
using Esri.ArcGISRuntime.Data;
using Esri.ArcGISRuntime.Geometry;
using Esri.ArcGISRuntime.Mapping;
using Esri.ArcGISRuntime.UI;
using System.IO;
//using System.Windows.Media;
```

生成全局变量，用于存储 shp 文件路径：

```csharp
string _file = @"ArcGISRuntimeSampleData\ChinaProvince\bou2_4p.shp";
```

编写 Initialize 函数：

```csharp
void Initialize()
{
    string filepath = System.IO.Path.Combine(AppDomain.CurrentDomain.BaseDirectory + "..\\..\\..\\..\\", _file);
    tbFile.Text = filepath;
    mapView1.GeoViewDoubleTapped += (s, e) =>
```

```
    {
        if (mapView1.SketchEditor.CompleteCommand.CanExecute(null))
            e.Handled=true;
    };
}
```

在构造函数中调用 Initialize 函数:
```
public MainWindow()
{
    InitializeComponent();
    Initialize();
}
```

打开地图:
```
async void btOpen_Click(object sender, RoutedEventArgs e)
{
    string filepath = tbFile.Text;
    if (!File.Exists(filepath))
        return;
    ShapefileFeatureTable sft = await ShapefileFeatureTable.OpenAsync(filepath);
    //Create a feature layer to display the shapefile.
    FeatureLayer fl = new FeatureLayer(sft);
    await fl.LoadAsync();
    mapView1.Map.OperationalLayers.Add(fl);
    mapView1.SetViewpoint(new Viewpoint(fl.FullExtent));
    QueryParameters qp = new QueryParameters();
    qp.WhereClause="1=1";
    FeatureQueryResult fqr = await sft.QueryFeaturesAsync(qp);
    Feature feature = fqr.FirstOrDefault();
    if (feature == null)
        return;
    fl.SelectFeature(feature);
}
```

编辑多边形:
```
async void btEdit_Click(object sender, RoutedEventArgs e)
```

```
}
    Geometry geometry = await mapView1.SketchEditor.StartAsync
(SketchCreationMode.Polygon, false);
    MapPoint mp = geometry as MapPoint;
    FeatureLayer fl = mapView1.Map.OperationalLayers[0] as Feature
Layer;
    FeatureQueryResult fqr = await fl.GetSelectedFeaturesAsync();
    Feature feature = fqr.FirstOrDefault();
    if (feature == null)
        return;
    feature.Geometry = geometry;
    ShapefileFeatureTable sft = fl.FeatureTable as ShapefileFeature
Table;
    await sft.UpdateFeatureAsync(feature);
}
```

第 16 节　要素集合图层查询器

要素集合图层被设计为显示中等数量的要素数据(数百个至数千个要素)，是在不同客户端共享静态数据的理想方式。要素信息在本地地理数据库(geodatabase)中缓存，在本地绘制，因此在地图移动或缩放时提供了优秀的显示性能。

【实验目的】

读取门户的要素集合图层，添加至业务图层进行显示。

【实验数据】

ArcGIS Server sample web service 在线地图：

http：//sampleserver6.arcgisonline.com/arcgis/rest/services/Wildfire/FeatureServer/0

【实验步骤】

1. 新建项目

.NET 框架：与 ArcGIS Runtime SDK 对应

模板列表：Visual C#/Windows/Windows Classic Desktop

模板：ArcGIS Runtime Application（WPF）

名称：FeatureCollectionLayerQueryer

位置：D：\ArcGISRuntimeTutorial

2. 编译项目

详情见第 1 章第 3 节。

3. 修改 MainWindow.xaml 标记语言

添加名称属性：

```
<Grid>
    <esri:MapView Name = "mapView1" Map = "{Binding Map, Source = {StaticResource MapViewModel}}" />
</Grid>
```

4. 修改 MainWindow.xaml.cs 代码

添加命名空间：

```
using Esri.ArcGISRuntime.Data;
using Esri.ArcGISRuntime.Mapping;
```

编写 Init 函数，其逻辑流程为：
ServiceFeatureTable→FeatureQueryResult→FeatureCollectionTable→FeatureCollection→FeatureCollectionLayer

```
void Init()
{
    string _FeatureTableUrl = "http://sampleserver6.arcgisonline.com/arcgis/rest/services/Wildfire/FeatureServer/0";
    string where = "1=1";
    Uri uri = new Uri(_FeatureTableUrl);
    //Create a service feature table to get features from.
    ServiceFeatureTable featTable = new ServiceFeatureTable(uri);
    //Create a query to get all features in the table.
    QueryParameters queryParams = new QueryParameters();
    queryParams.WhereClause = where;
    //Query the table to get all features.
    FeatureQueryResult queryResult = await featTable.QueryFeaturesAsync(queryParams);
    //Create a new feature collection table from the result features
    FeatureCollectionTable collectTable = new FeatureCollectionTable(queryResult);
    //Create a feature collection and add the table.
    FeatureCollection featCollection = new FeatureCollection();
    featCollection.Tables.Add(collectTable);
    //Create a layer to display the feature collection.
    FeatureCollectionLayer featCollectionTable = new FeatureCollectionLayer(featCollection);
```

```
    mapView1.Map.OperationalLayers.Add(featCollectionTable);
}
```
在构造函数中调用 Init 函数：
```
public MainWindow()
{
    InitializeComponent();
    Init();
}
```
【实验结果】

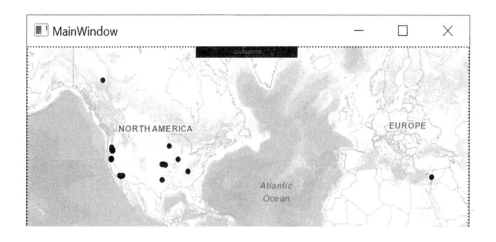

第 17 节　要素集合图层构建器

要素集合图层(FeatureCollectionLayer)用来显示来自要素集合表中的要素。集合中的要素具有不同的构架（模式，scheme），几何类型和渲染方式。创建要素表格(FeatureCollectionTable)可分为三步：（1）常见表结构；（2）添加要素，填充属性信息和几何信息；（3）设置要素显示时的符号系统。

【实验目的】

通过多个要素集表格，生成要素集，作为要素集合图层在地图中显示。

【实验数据】

中国四个大城市的经纬度：

　　　　{116.41667, 39.91667}, //北京
　　　　{113.00000, 28.21667}, //长沙
　　　　{113.23333, 23.16667}, //广州
　　　　{108.37, 21.57}//钦州

第 17 节　要素集合图层构建器

【实验步骤】

1. 新建项目

.NET 框架：与 ArcGIS Runtime SDK 对应
模板列表：Visual C#/Windows/Windows Classic Desktop
模板：ArcGIS Runtime Application（WPF）
名称：FeatureCollectionLayer
位置：D：\ArcGISRuntimeTutorial

2. 编译项目

详情见第 1 章第 3 节。

3. 修改 MainWindow.xaml 标记语言

```
<Grid>
    <esri:MapView Name="mapView1" Map="{Binding Map,Source={StaticResource MapViewModel}}">
</Grid>
```

4. 修改 MainWindow.xaml.cs 代码

添加命名空间：

```
using Esri.ArcGISRuntime.Data;
using Esri.ArcGISRuntime.Geometry;
using Esri.ArcGISRuntime.Mapping;
using Esri.ArcGISRuntime.Symbology;
```

用数组存储四个城市的坐标：

```
double[,] _points = {
    {116.41667,39.91667},//北京
    {113.00000,28.21667},//长沙
    {113.23333,23.16667},//广州
    {108.37,21.57}//钦州
};
```

编写 CreatePointTable 函数，添加点表格：

```
FeatureCollectionTable CreatePointTable()
{
    Field field = new Field(FieldType.Text, "Place", "place name", 50);
    Field[] fields = { field };
    FeatureCollectionTable table = new FeatureCollectionTable
```

```
(fields, GeometryType.Point, SpatialReferences.Wgs84);
    MapPoint mp0 = new MapPoint(_points[0, 0], _points[0, 1], Spatial
References.Wgs84);
    Feature feature = table.CreateFeature();
    feature.SetAttributeValue(field, "Beijing");
    feature.Geometry = mp0;
    table.AddFeatureAsync(feature);
    Symbol sym = new SimpleMarkerSymbol(SimpleMarkerSymbolStyle.
Triangle, Colors.Red, 18);
    table.Renderer = new SimpleRenderer(sym);
    return table;
}
```

编写函数, 添加线表格:

```
FeatureCollectionTable CreateLineTable()
{
    Field field = new Field(FieldType.Text, "boundary", "boundary
name", 50);
    Field[] fields = { field };
    FeatureCollectionTable table = new FeatureCollectionTable
(fields, GeometryType.Polyline, SpatialReferences.Wgs84);
    MapPoint mp0 = new MapPoint(_points[0, 0], _points[0, 1],
SpatialReferences.Wgs84);
    MapPoint mp1 = new MapPoint(_points[1, 0], _points[1, 1],
SpatialReferences.Wgs84);
    MapPoint[] mps = { mp0, mp1 };
    var line = new Esri.ArcGISRuntime.Geometry.Polyline(mps);
    Feature feature = table.CreateFeature();
    feature.SetAttributeValue(field, "railway");
    feature.Geometry = line;
    table.AddFeatureAsync(feature);
    Symbol sym = new SimpleLineSymbol(SimpleLineSymbolStyle.Dash,
Colors.Green, 3);
    table.Renderer = new SimpleRenderer(sym);
    return table;
}
```

编写函数, 添加多边形表格:

```
FeatureCollectionTable CreateGonTable()
{
```

```
    Field field = new Field(FieldType.Text, "area", "area name", 50);
    Field[] fields = { field };
       FeatureCollectionTable  table  =  new  FeatureCollectionTable
(fields, GeometryType.Polygon, SpatialReferences.Wgs84);
    Feature feature = table.CreateFeature();
    feature.SetAttributeValue(field, "triangle");
    MapPoint mp1 = new MapPoint(_points[1, 0], _points[1, 1], Spatial
References.Wgs84);
    MapPoint mp2 = new MapPoint(_points[2, 0], _points[2, 1], Spatial
References.Wgs84);
    MapPoint mp3 = new MapPoint(_points[3, 0], _points[3, 1], Spatial
References.Wgs84);
    MapPoint[] mps = { mp1, mp2, mp3 };
    var gon = new Esri.ArcGISRuntime.Geometry.Polygon(mps);
    feature.Geometry = gon;
    table.AddFeatureAsync(feature);
    SimpleLineSymbol lineSym = new SimpleLineSymbol(SimpleLineSymbol
Style.Solid, Colors.DarkBlue, 2);
    Symbol sym = new SimpleFillSymbol(SimpleFillSymbolStyle.Diagonal
Cross, Colors.Cyan, lineSym);
    table.Renderer = new SimpleRenderer(sym);
    return table;
}
```

编写函数，通过三个表格生成要素集合图层：

```
FeatureCollectionLayer CreateLayer()
{
    FeatureCollection fc = new FeatureCollection();
    FeatureCollectionTable pointTable = CreatePointTable();
    FeatureCollectionTable lineTable = CreateLineTable();
    FeatureCollectionTable gonTable = CreateGonTable();
    fc.Tables.Add(pointTable);
    fc.Tables.Add(lineTable);
    fc.Tables.Add(gonTable);
    FeatureCollectionLayer layer = new FeatureCollectionLayer(fc);
    return layer;
}
```

编写 Init 函数，设置观察视点：

```
void Init()
```

```
{
    FeatureCollectionLayer layer = CreateLayer();
    await layer.LoadAsync();
    await mapView1.SetViewpointAsync(new Viewpoint(layer.FullExtent));
    mapView1.Map.OperationalLayers.Add(layer);
}
```
在构造函数中调用 Init 函数：
```
public MainWindow()
{
    InitializeComponent();
    Init();
}
```
【实验结果】

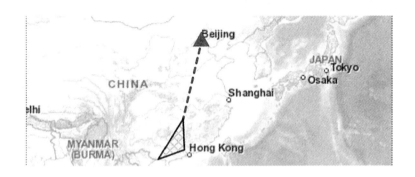

第6章 实用工具篇

ArcGIS Runtime 具有轻量级、绿色免安装特点，API 丰富、功能强大，支持离线和在线应用，因此非常适合制作即拿即用的工具。

由于 API 独立，具有较好的跨平台特性，因此可以利用 .NET 丰富的计算生态，集成第三方 SDK，从而快速构建通用型或专用型 App。

第1节 地图坐标查看器

目前网络电子地图(包括导航)主要使用的坐标系叫网络墨卡托(Web Mercator)，是 Google 发明的，在 Google Map 最先使用。Web Mercator 坐标系是一个被 EPSG(European Petroleum Survey Group)称为伪墨卡托投影(Popular Visualization Pseudo Mercator，PVPM)。在投影过程中，将表示地球的参考椭球体近似的作为正球体处理(正球体半径 R = 椭球体半长轴 a)。在 ArcGIS 中这个坐标系叫 WGS 1984 Web Mercator (Auxiliary Sphere)。EPSG 确定为 WKID：3857，对应于 Esri 内部使用 ID ESRI：102100。

【实验目的】

在地图上移动鼠标时，实时显示鼠标所在点的地图坐标。

【实验数据】

ArcGIS 在线底图服务，Runtime App 的默认地图。

【实验步骤】

1. 新建项目

.NET 框架：与 ArcGIS Runtime SDK 对应

模板列表：Visual C#/Windows/Windows Classic Desktop

模板：ArcGIS Runtime Application (WPF)

名称：CoordViewer

位置：D:\ArcGISRuntimeTutorial

2. 编译项目

详情见第1章第3节。

3. 修改 MainWindow.xaml 标记语言

对地图视图添加名称属性，并添加鼠标移动事件：

```xml
<esri:MapView Name="mapView1" Map="{Binding Map,Source={StaticResource MapViewModel}}" MouseMove="MapView_MouseMove"/>
```

在 App 下部添加状态栏，放入文本框，用于显示坐标：

```xml
<StatusBar VerticalAlignment="Bottom" HorizontalAlignment="Left">
<TextBox Name="tbCoord"/>
</StatusBar>
```

4. 修改 MainWindow.xaml.cs 代码

添加命名空间：

```csharp
using Esri.ArcGISRuntime.Mapping;
using Esri.ArcGISRuntime.Geometry;
```

编写鼠标移动事件响应函数：

```csharp
private void MapView_MouseMove(object sender, MouseEventArgs e)
{
Point cursorSceenPoint = e.GetPosition(mapView1);
//Get the corresponding MapPoint.
MapPoint mp = mapView1.ScreenToLocation(cursorSceenPoint);
string mpDesc = string.Empty;
// Return if the MapPoint is null. This might happen if mouse leaves MapView area.
if (mp != null)
mpDesc = string.Format("Long:{1:F6},Lat:{0:F6}", mp.X, mp.Y);
tbCoord.Text = mpDesc;
}
```

【实验结果】

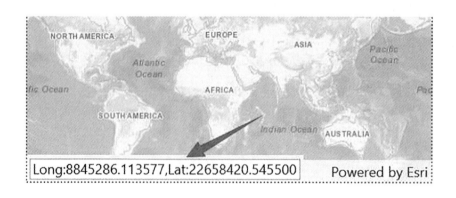

第 2 节　图层视图状态查看器

图层视图状态描述图层是否可见、是否激活。在对图层进行查询、分析和编辑前，需要先查询图层的状态，从而确定当前操作是否有效。

【实验目的】

读取图层名称和图层状态；当图层状态改变时，通过委托进行通知。

【实验数据】

ArcGIS 在线底图服务，Runtime App 的默认地图。

【实验步骤】

1. 新建项目

.NET 框架：与 ArcGIS Runtime SDK 对应

模板列表：Visual C#/Windows/Windows Classic Desktop

模板：ArcGIS Runtime Application（WPF）

名称：LayerViewStateViewer

位置：D：\ArcGISRuntimeTutorial

2. 编译项目

详情见第 1 章第 3 节。

3. 修改 MainWindow.xaml 标记语言

添加 TextBox，用于存储图层名称和状态：

```
<esri:MapView Name="mapView1" />
<TextBox Name="tbLayerViewState" Width="200" Height="200" HorizontalAlignment="Left" VerticalAlignment="Top" Opacity="0.6"/>
```

4. 修改 MainWindow.xaml.cs 代码

```csharp
public void Init()
{
    mapView1.LayerViewStateChanged += (s, e) =>
    {
        string lyrName = e.Layer.Name;
        tbLayerViewState.Text += lyrName + ": " + e.LayerViewState.Status.ToString() + Environment.NewLine;
    };
}
```

```
public MainWindow()
{
    InitializeComponent();
    Init();
}
```
【实验结果】

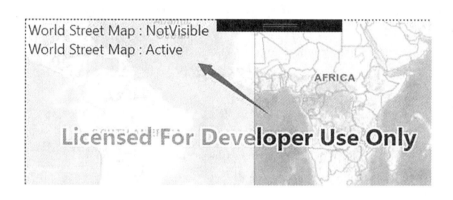

第3节　坐标投影转换器

地理坐标系是在球体上定义的角度坐标系。其中 GCS_WGS_1984（世界大地坐标系 1984，WGS1984）是 GPS 采用的坐标系，Wkid（Well-known ID）= 4326。WGS1984 是美国国防部制图局（Defence Mapping Agency，DMA）为统一世界大地坐标系统，实现全球测量标准的一致性，而定义的用于制图、大地测量、导航的坐标基准，包括标准地球坐标框架、用于处理原始观测数据的标准椭球参考面（即基准和参考椭球）和定义标准海平面的重力等势面（大地水准面）。在地理坐标系和投影坐标系间相互转换，是地图数据处理的常见任务。

【实验目的】

在地图上移动鼠标时，实时读取鼠标所在点的地图坐标。ArcGIS 在线地图服务的坐标一般为 Web Mercator 坐标系，将其转换为 WGS1984 坐标系。

【实验数据】

ArcGIS 在线底图服务，Runtime App 的默认地图。

【实验步骤】

1. 新建项目

.NET 框架：与 ArcGIS Runtime SDK 对应

模板列表：Visual C#/Windows/Windows Classic Desktop

模板：ArcGIS Runtime Application（WPF）

名称：CoordProjecter

位置：D：\ArcGISRuntimeTutorial

2. 编译项目

详情见第 1 章第 3 节。

3. 修改 MainWindow. xaml 标记语言

对地图视图添加名称属性，并添加鼠标移动事件：

```
<esri:MapView Name="mapView1" Map="{Binding Map, Source={Static
Resource MapViewModel}}" MouseMove="mapView1_MouseMove" />
```

在 App 下部添加状态栏，放入文本框，用于显示坐标：

```
<StatusBar VerticalAlignment="Bottom" HorizontalAlignment="Left">
<TextBox Name="tbCoord" />
</StatusBar>
```

4. 修改 MainWindow. xaml. cs 代码

添加命名空间：

```
using Esri.ArcGISRuntime.Mapping;
using Esri.ArcGISRuntime.Geometry;
```

编写鼠标移动事件响应函数：

```
private void mapView1_MouseMove(object sender, MouseEventArgs e)
{
Point cursorSceen = e.GetPosition(mapView1);
//Get the corresponding MapPoint.
MapPoint mp = mapView1.ScreenToLocation(cursorSceen);
// Return if the MapPoint is null. This might happen if mouse leaves
MapView area.
string mapLocDesc = string.Empty;
if (mp! = null)
    {
        MapPoint projected = (MapPoint)GeometryEngine.Project(mp,
SpatialReferences.Wgs84);
        mapLocDesc = string.Format("Long:{1:F6},Lat:{0:F6}",
projected.X, projected.Y);
}
tbCoord.Text = mapLocDesc;
}
```

【实验结果】

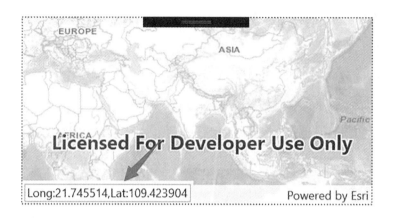

第 4 节　shp 工作空间读取器

工作空间一般是一组满足特定要求的数据和资源的集合。例如一个 GDB、MDB 数据库，或者一个数据集。由于 SHPs 是本地文件，因此多个 SHP 数据组成的文件夹，可以认为就是一个工作空间。

【实验目的】

读取某个文件夹的所有 SHP 数据。

【实验数据】

ArcGIS Online 在线底图服务：
https:// sampleserver6.arcgisonline.com/ arcgis/ rest/ services/ World_Street_Map/MapServer

本地 shp 文件，要素内容为小比例尺的中国各省级行政区划边界：ChinaProvince\bou2_4p.shp

【实验步骤】

1. 新建项目

.NET 框架：与 ArcGIS Runtime SDK 对应

模板列表：Visual C#/Windows/Windows Classic Desktop

模板：ArcGIS Runtime Application（WPF）

名称：ShapefileWorkspaceReader

位置：D：\ArcGISRuntimeTutorial

2. 编译项目

详情见第 1 章第 3 节。

3. 修改 MainWindow.xaml 标记语言

对地图视图,添加名称属性:
```
<esri:MapView Name="mapView1" Map="{Binding Map,Source={Static
Resource MapViewModel}}"/>
```
设计 UI:
```
<StackPanel Opacity="0.5">
    <TextBox Name="tbWorkspace"/>
    <Button Name="btOpen" Content="Open" Click="btOpen_Click"/>
    <TextBox Name="tbFiles"/>
</StackPanel>
```

4. 修改 MainWindow.xaml.cs 代码

添加命名空间:
```
using Esri.ArcGISRuntime.Data;
using Esri.ArcGISRuntime.Geometry;
using Esri.ArcGISRuntime.Mapping;
```
生成全局变量,用于存储 shp 文件路径:
```
string _file = @"ArcGISRuntimeSampleData\China Provinces\";
```
编写 Initialize 函数:
```
void Initialize()
{
    string filepath = System.IO.Path.Combine(AppDomain.CurrentDomain.BaseDirectory + "..\\..\\..\\..\\", _file);
    tbWorkspace.Text = filepath;
}
```
在构造函数中调用 Initialize 函数:
```
public MainWindow()
{
    InitializeComponent();
    Initialize();
}
```
点击打开按钮,读取工作空间下所有 *.shp 文件,通过图层,加入到地图视图中,同时记录图层名称(Name)和当前范围(FullExtent):
```
async void btOpen_Click(object sender, RoutedEventArgs e)
{
    string filepath = tbWorkspace.Text.Trim();
    string[] files = System.IO.Directory.GetFiles(filepath, "*
```

```
.shp");
    List<Envelope> extents = new List<Envelope>();
    foreach (string file in files)
    {
        ShapefileFeatureTable sft = await ShapefileFeatureTable.OpenAsync(file);
        //Create a feature layer to display the shapefile.
        FeatureLayer fl = new FeatureLayer(sft);
        await fl.LoadAsync();
        mapView1.Map.OperationalLayers.Add(fl);
        //Add the extent to the list of extents.
        extents.Add(fl.FullExtent);
        tbFiles.Text += fl.Name + Environment.NewLine;
    }
    // Use the geometry engine to compute the full extent of the ENC Exchange Set.
    Envelope fullExtent = GeometryEngine.CombineExtents(extents);
    //Set the viewpoint.
    mapView1.SetViewpoint(new Viewpoint(fullExtent));
}
```

【实验结果】

第5节 地名标注器

在地图上进行名称、类型标注，是信息采集的常用方式。

【实验目的】

交互标注：用户在文本框输入标注内容，用鼠标点击标注位置，进行动态标注。

【实验数据】

ArcGIS 在线底图服务，Runtime App 的默认地图。

【实验步骤】

1. 新建项目

.NET 框架：与 ArcGIS Runtime SDK 对应
模板列表：Visual C#/Windows/Windows Classic Desktop
模板：ArcGIS Runtime Application（WPF）
名称：LocationLabeler
位置：D：\ArcGISRuntimeTutorial

2. 编译项目

详情见第 1 章第 3 节。

3. 修改 MainWindow.xaml 标记语言

对地图视图，添加名称属性和鼠标事件：

```
<esri:MapView Name="mapView1" Map="{Binding Map, Source={Static
Resource MapViewModel}}" MouseDown="mapView1_MouseDown" />
```

添加工具条，文本框子控件输入标注内容，复选框指示标注开始/结束状态：

```
<ToolBar HorizontalAlignment="Left" VerticalAlignment="Top"
Opacity="0.9">
<TextBox Name="tbLabel" Text="北部湾大学" MinWidth="100" />
<CheckBox Name="cbLabel" Content="Label" IsChecked="True" />
</ToolBar>
```

4. 修改 MainWindow.xaml.cs 代码

添加命名空间：

```
using Esri.ArcGISRuntime.Mapping;
using Esri.ArcGISRuntime.Geometry;
```

编写鼠标按下事件，处理用户鼠标右键点击行为，其逻辑为通过复选框判断当前标注状态→判断鼠标右键状态→读取鼠标点击位置→读取文本框内容→生成 callout 并显示：

```
private void mapView1_MouseDown(object sender, MouseButtonEventArgs e)
{
if (! cbLabel.IsChecked.Value)
    return;
if (e.RightButton == MouseButtonState.Pressed)
  {
    Point cursorSceen = e.GetPosition(mapView1);
    //Get the corresponding MapPoint.
```

```
    MapPoint mp = mapView1.ScreenToLocation(cursorSceen);
    //Return if the MapPoint is null. This might happen if mouse leaves
MapView area.
    string text = tbLabel.Text.Trim();
     CalloutDefinition callout = new CalloutDefinition("Label:",
text);
    mapView1.ShowCalloutAt(mp, callout);
  }
}
```

【实验结果】

使用方法：左上角文本框输入标注内容，点击 Label 按钮切换（启用或者关闭）标注功能，右键点击地图即在对应位置放置标注。

第 6 节　设备地址显示

大部分设备，包括台式电脑，笔记本，平板电脑和手机等，都支持定位信息。具体技术手段包括 IP 地址、Wi-FI、移动网络（蜂窝网络，cellular networks）、GPS 或者北斗等卫星定位系统等。因此有必要显示运行设备的实时地址。

【实验目的】

显示运行设备实时地址。由于内置支持定位功能，因此可以支持台式电脑、笔记本、平板电脑、手机等进行定位，即使在没有 GPS(GNSS) 的条件下，也可以通过多种辅助手段（例如 IP 定位）进行定位。

【实验数据】

ArcGIS Online 在线底图服务：

https://sampleserver6.arcgisonline.com/arcgis/rest/services/World_Street_Map/MapServer

矢量数据：本地 shp 文件。要素内容为小比例尺的中国各省级行政区划边界：ChinaProvince \ bou2_4p.shp

第 6 节 设备地址显示

【实验步骤】

1. 新建项目

.NET 框架：与 ArcGIS Runtime SDK 对应
模板列表：Visual C#/Windows/Windows Classic Desktop
模板：ArcGIS Runtime Application（WPF）
名称：DeviceLocationDisplay
位置：D：\ArcGISRuntimeTutorial

2. 编译项目

详情见第 1 章第 3 节。

3. 修改 MainWindow.xaml 标记语言

对地图视图，添加名称属性：

```
<Grid>
    <esri:MapView Name="mapView1" Map="{Binding Map,Source={StaticResource MapViewModel}}" />
</Grid>
```

设计 UI，添加一个堆叠面板，子控件包括用于输入的复选框和按钮：

复选框：提供设备地址显示模式，供用户选择。

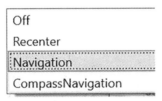

按钮：显示或隐藏设备地址显示功能。

```
<StackPanel Opacity="0.8">
    < ComboBox Name = " cbLocationDisplayModes " SelectionChanged = "cbLocationDisplayModes_SelectionChanged" Opacity="0.8" />
    <Button Content="Show LocationDisplay" Name="ShowLocation
```

```xml
Display" Click="ShowLocationDisplay_Click"/>
</StackPanel>
```

4. 修改 MainWindow.xaml.cs 代码

添加命名空间：

```csharp
using Esri.ArcGISRuntime.Mapping;
```

编写 Initialize 函数：

```csharp
void Initialize()
{
    //Set navigation types as items source and set default value.
    cbLocationDisplayModes.ItemsSource = Enum.GetValues(typeof(LocationDisplayAutoPanMode));//_locationDisplayModes;
    cbLocationDisplayModes.SelectedIndex = 0;
    mapView1.LocationDisplay.IsEnabled = !mapView1.LocationDisplay.IsEnabled;
}
```

在构造函数中调用 Initialize 函数：

```csharp
public MainWindow()
{
    InitializeComponent();
    Initialize();
}
```

用户点击复选框，选择设备地址显示模式：

```csharp
private void cbLocationDisplayModes_SelectionChanged(object sender, SelectionChangedEventArgs e)
{
    mapView1.LocationDisplay.AutoPanMode = (LocationDisplayAutoPanMode)cbLocationDisplayModes.SelectedItem;
}
```

启用或者关闭设备地址显示功能：

```csharp
private void ShowLocationDisplay_Click(object sender, RoutedEventArgs e)
{
    mapView1.LocationDisplay.IsEnabled = !mapView1.LocationDisplay.IsEnabled;
}
```

【实验结果】

第 7 节　设备地址坐标读取器

大部分设备，包括台式电脑，笔记本，平板电脑和手机等，都能提供定位信息。这些平台提供 API 接口，通过如 Wi-FI，移动网络(蜂窝网络，cellular networks)，GPS 或者北斗等卫星定位系统获取位置信息。

【实验目的】

显示 App 运行的设备所在的地理坐标。

【实验数据】

ArcGIS Online 在线底图服务：

https：//sampleserver6.arcgisonline.com/arcgis/rest/services/World_Street_Map/MapServer

矢量数据：本地 shp 文件，内容为小比例尺的中国各省级行政区划边界：ChinaProvince \ bou2_4p.shp

【实验步骤】

1. 新建项目

.NET 框架：与 ArcGIS Runtime SDK 对应
模板列表：Visual C#/Windows/Windows Classic Desktop
模板：ArcGIS Runtime Application（WPF）
名称：DeviceLocationCoord
位置：D：\ArcGISRuntimeTutorial

2. 编译项目

详情见第 1 章第 3 节。

3. 修改 MainWindow.xaml 标记语言

对地图视图,添加名称属性:
```
<Grid>
    <esri:MapView Name="mapView1" Map="{Binding Map, Source={StaticResource MapViewModel}}" />
</Grid>
```
设计 UI,添加一个文本框,用于显示设备当前实时坐标:
```
<TextBox Name="tblocation" HorizontalAlignment="Left" VerticalAlignment="Top" Opacity="0.8" />
```

4. 修改 MainWindow.xaml.cs 代码

添加命名空间:
```
using Esri.ArcGISRuntime.Data;
using Esri.ArcGISRuntime.Mapping;
using System.IO;
```
生成全局变量,用于存储 shp 文件路径:
```
string _shpPath = @"ChinaProvince\bou2_4p.shp";
```
编写 Initialize 函数:
```
void Initialize()
{
    //Get the path to the shapefile.
    string filepath = System.IO.Path.Combine(AppDomain.CurrentDomain.BaseDirectory, _shpPath);
    if (!File.Exists(filepath))
        return;
    //Open the shapefile.
    ShapefileFeatureTable myShapefile = await ShapefileFeatureTable.OpenAsync(filepath);
    //Create a feature layer to display the shapefile.
    FeatureLayer newFeatureLayer = new FeatureLayer(myShapefile);
    //Add the feature layer to the map.
    mapView1.Map.OperationalLayers.Add(newFeatureLayer);
    //Zoom the map to the extent of the shapefile.
    await mapView1.SetViewpointGeometryAsync(newFeatureLayer.FullExtent);
    mapView1.LocationDisplay.IsEnabled = !mapView1.Location
```

```
Display.IsEnabled;
    mapView1.LocationDisplay.LocationChanged + = LocationDisplay_
LocationChanged;
```
在构造函数中调用 Initialize 函数：
```
public MainWindow()
{
    InitializeComponent();
    Initialize ();
}
```
设备坐标变化响应事件：
```
private void LocationDisplay_LocationChanged(object sender,Esri.
ArcGISRuntime.Location.Location e)
{
    string location = (e == null ||e.Position == null) ? string.
Empty: e.Position.ToString();
    string mapLocation = string.Empty;
    Dispatcher.Invoke(() =>
    {
        //mapLocation = mapView1.LocationDisplay.MapLocation.
ToString();
        tblocation.Text = location;
    });
}
```

第 8 节　进度查看器

在加载耗时资源(如空间数据)、计算密集型任务(如深度学习)等时，进度条让用户了解长时间暂停是因为卡死(资源被锁死、网络链接断开)，还是正在运行，从而提高用户参与感，提升 App 体验感。如果没有进度条，用户可能在中途因为等待时间过长而强制退出。

【实验目的】

在地图加载过程中，使用动态显示进度条；资源加载完成后，隐藏进度条。

【实验数据】

ArcGIS 在线底图服务，Runtime App 的默认地图。

【实验步骤】

1. 新建项目

.NET 框架：与 ArcGIS Runtime SDK 对应

模板列表：Visual C#/Windows/Windows Classic Desktop
模板：ArcGIS Runtime Application（WPF）
名称：DrawStatusViewer
位置：D：\ArcGISRuntimeTutorial

2. 编译项目

详情见第1章第3节。

3. 修改 MainWindow.xaml 标记语言

```
<ProgressBar Name="pbDrawProgress" Height="10" IsIndeterminate="True" Opacity="0.5"/>
```

4. 修改 MainWindow.xaml.cs 代码

```csharp
public MainWindow()
{
    InitializeComponent();
    Init();
}
public void Init()
{
    mapView1.DrawStatusChanged += (s, e) =>
    {
        if (e.Status == DrawStatus.InProgress)
            pbDrawProgress.Visibility = Visibility.Visible;
        else if (e.Status == DrawStatus.Completed)
            pbDrawProgress.Visibility = Visibility.Hidden;
    };
}
```

【实验结果】

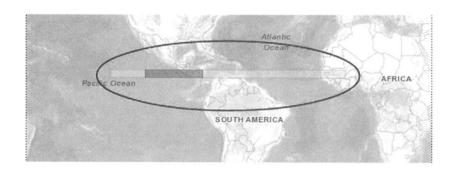

第 9 节　服务要素表手动缓存管理器

针对网络服务，对资源提前进行下载和缓存，可以较大提升系统性能和用户体验。手动进行缓存，可以增加系统的灵活性。

【实验目的】

通过手动，对目标区域要素进行缓存。

ServiceFeatureTable.FeatureRequestMode = ManualCache 设置为手动缓存模式。

【实验数据】

ArcGIS Server sample web service 在线地图。

【实验步骤】

1. 新建项目

.NET 框架：与 ArcGIS Runtime SDK 对应

模板列表：Visual C#/Windows/Windows Classic Desktop

模板：ArcGIS Runtime Application（WPF）

名称：ServiceFeatureTableCacheManager

位置：D：\ArcGISRuntimeTutorial

2. 编译项目

详情见第 1 章第 3 节。

3. 修改 MainWindow.xaml 标记语言

添加名称属性：

```
<Grid>
    <esri:MapView Name="mapView1" Map="{Binding Map,Source={StaticResource MapViewModel}}" />
</Grid>
```

4. 修改 MainWindow.xaml.cs 代码

添加命名空间：

```
using Esri.ArcGISRuntime;
using Esri.ArcGISRuntime.Data;
using Esri.ArcGISRuntime.Geometry;
using Esri.ArcGISRuntime.Mapping;
```

定义全局变量：

```
string _url = "http://sampleserver6.arcgisonline.com/arcgis/rest/services/SF311/FeatureServer/0";
```

```
ServiceFeatureTable _incidentsFeatureTable;
```
编写 Initialize 函数：
```
void Initialize()
{
    var serviceUri = new Uri(_url);
    //为要素服务生成要素表
    _incidentsFeatureTable = new ServiceFeatureTable(serviceUri);
    //定义请求模式
    _incidentsFeatureTable.FeatureRequestMode = FeatureRequestMode.ManualCache;
    //数据表加载后填充数据
    _incidentsFeatureTable.LoadStatusChanged += FeatureTable_LoadStatusChanged;
    //通过要素表生成要素图层
    FeatureLayer incidentsFeatureLayer = new FeatureLayer(_incidentsFeatureTable);
    mapView1.Map.OperationalLayers.Add(incidentsFeatureLayer);
}
```
编写缓存函数，在要素表格加载完后，通过查询手动缓存：
```
async void FeatureTable_LoadStatusChanged(object sender, LoadStatusEventArgs e)
{
    //图层未成功加载，返回
    if (e.Status != LoadStatus.Loaded)
        return;
    //通过查询参数生成查询对象
    QueryParameters queryParameters = new QueryParameters()
    {
        WhereClause = "req_Type = 'Tree Maintenance or Damage'"
    };
    //返回字段
    var outputFields = new string[] { "*" };
    //查询返回数据填充值要素表
    await _incidentsFeatureTable.PopulateFromServiceAsync(queryParameters, true, outputFields);
}
```
在构造函数中调用 Initialize 函数：
```
public MainWindow()
```

```
{
    InitializeComponent();
    Initialize();
}
```
【实验结果】

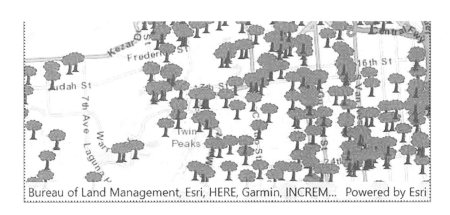

第 10 节　移动地理数据库浏览器

移动地理数据库(mobile geodatabase)，又叫运行时地理数据库(runtime geodatabase)，扩展名为(.geodatabase)，是不同类型的数据集的容器，基于移动 SQLite 格式，主要用于 Runtime 离线工作流。在 windows 文件系统里，geodatabase 可达 1TB。

【实验目的】

用户输入移动地理数据库(.geodatabase)路径后，打开移动数据库，加载每个图层并显示出来。

【实验数据】

本地移动地理数据库，内容为中国省级行政区划及一级河流：

geodatabase \ provinceandriver.geodatabase

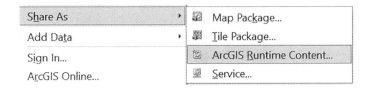

【实验步骤】

1. 新建项目

.NET 框架：与 ArcGIS Runtime SDK 对应

模板列表：Visual C#/Windows/Windows Classic Desktop
模板：ArcGIS Runtime Application（WPF）
名称：GeodatabaseViewer
位置：D：\ArcGISRuntimeTutorial

2. 编译项目

详情见第 1 章第 3 节。

3. 修改 MainWindow.xaml 标记语言

对地图视图添加名称属性：
```
<esri:MapView Name="mapView1" Map="{Binding Map, Source={StaticResource MapViewModel}}"/>
```
设计 UI，添加一个堆叠面板，子控件包括用于输入 geodatabase 路径的文本框和加载 geodatabase 的按钮：
```
<StackPanel Background="Wheat" Opacity="0.9" Margin="3" Orientation="Horizontal" HorizontalAlignment="Left" VerticalAlignment="Top">
    <TextBox Name="tbGeodatabase" Text="geodatabase\provincean driver.geodatabase" Margin="3" Padding="3" MinWidth="200"/>
    <Button Name="btLoadGeodatabase" Content="LoadGeodatabase" Padding="3" Margin="3" Click="btLoadGeodatabase_Click"/>
</StackPanel>
```

4. 修改 MainWindow.xaml.cs 代码

添加命名空间：
```
using Esri.ArcGISRuntime.Data;
using Esri.ArcGISRuntime.Geometry;
using Esri.ArcGISRuntime.Mapping;
```
编写代码响应用户点击事件，加载 geodatabase 中的每个图层：
```
async void btLoadGeodatabase_Click(object sender, RoutedEventArgs e)
{
    string geodatabasePath = Path.Combine(AppDomain.CurrentDomain.BaseDirectory, tbGeodatabase.Text.Trim());
    //Get the new geodatabase.
    Geodatabase gdb = await Geodatabase.OpenAsync(geodatabasePath);
    Viewpoint vp = null;
    //Loop through all feature tables in the geodatabase.
    foreach (GeodatabaseFeatureTable table in gdb.GeodatabaseFeatureTables)
```

```
    }
        //Create a new feature layer for the table.
        FeatureLayer layer = new FeatureLayer(table);
        await layer.LoadAsync();
        if (vp == null)
        {
            vp = new Viewpoint(layer.FullExtent);
            mapView1.SetViewpoint(vp);
        }
        mapView1.Map.OperationalLayers.Add(layer);
    }
}
```

第 11 节　移动地理数据库下载器

对于有下载权限的在线地图数据服务，在空余时间提前下载，可以较好提升用户体验。使用国外的数据集时，网速较慢，利用 VPN 可以显著提升国外网站下载速度。

【实验目的】

创建本地数据库，异步下载服务资源。

本节实验内容较多，逻辑非常复杂。读者需要反复分析代码，弄懂技术路线，厘清数据流程。

【实验数据】

ArcGIS Online 在线矢量要素服务，内容为野火分布：

https：//sampleserver6.arcgisonline.com/arcgis/rest/services/Sync/WildfireSync/FeatureServer

【实验步骤】

1. 新建项目

.NET 框架：与 ArcGIS Runtime SDK 对应

模板列表：Visual C#/Windows/Windows Classic Desktop

模板：ArcGIS Runtime Application（WPF）

名称：GeodatabaseDownloader

位置：D：\ArcGISRuntimeTutorial

2. 编译项目

详情见第 1 章第 3 节。

3. 修改 MainWindow.xaml 标记语言

对地图视图添加名称属性：

<esri:MapView Name="mapView1" Map="{Binding Map, Source={StaticResource MapViewModel}}" />

设计 UI 如下，添加堆叠控件：

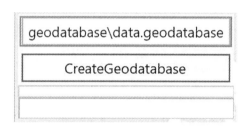

控件功能如下：

文本框：下载的 geodatabase 的路径

按钮：下载 geodatabase 命令

进度条：下载进度

文本框：下载消息列表

编写 xaml 代码如下：

```
<StackPanel Background="Wheat" Opacity="0.9" Margin="3" HorizontalAlignment="Left" VerticalAlignment="Top">
    <TextBox Name="tbGdb" Text="geodatabase\data.geodatabase" Margin="3" Padding="3" />
    <Button Name="btCreateGeodatabase" Content="CreateGeodatabase" Padding="3" Margin="3" Click="btCreateGeodatabase_Click" />
    <ProgressBar Name="progressBar1" MinHeight="10" Visibility="Collapsed" />
    <TextBox Name="tbMsg" VerticalScrollBarVisibility="Auto" />
</StackPanel>
```

4. 修改 MainWindow.xaml.cs 代码

添加命名空间：

```
using Esri.ArcGISRuntime.Data;
using Esri.ArcGISRuntime.Geometry;
using Esri.ArcGISRuntime.Mapping;
using Esri.ArcGISRuntime.Symbology;
```

```
using Esri.ArcGISRuntime.Tasks;
using Esri.ArcGISRuntime.Tasks.Offline;
using Esri.ArcGISRuntime.UI;
using Esri.ArcGISRuntime.UI.Controls;
```
定义全局变量,存储矢量要素服务:
```
Uri _featureServiceUri = new Uri("https://sampleserver6.arcgisonline.com/arcgis/rest/services/Sync/WildfireSync/FeatureServer");
```
编写地图初始化函数 InitMap:
```
async void InitMap()
{
    //Create a task for generating a geodatabase
    GeodatabaseSyncTask _gdbSyncTask = await GeodatabaseSyncTask.CreateAsync(_featureServiceUri);
    foreach (var layer in _gdbSyncTask.ServiceInfo.LayerInfos)
    {
        //Create the ServiceFeatureTable for this particular layer
        ServiceFeatureTable onlineTable = new ServiceFeatureTable(new Uri(_featureServiceUri + "/" + layer.Id));
        //Wait for the table to load
        await onlineTable.LoadAsync();
        // Add the layer to the map's operational layers if load succeeds
        if (onlineTable.LoadStatus == Esri.ArcGISRuntime.LoadStatus.Loaded)
        {
            mapView1.Map.OperationalLayers.Add(new FeatureLayer(onlineTable));
        }
    }
    Envelope enve = new Envelope(-122.53170503036097, 37.69818416320 2356, -122.35276956272253, 37.840951823552174, SpatialReferences.Wgs84);
    mapView1.SetViewpoint(new Viewpoint(enve));
    SimpleLineSymbol sls = new SimpleLineSymbol(SimpleLineSymbolStyle.DashDot, Colors.Red, 2);
    GraphicsOverlay go = new GraphicsOverlay()
```

```
        }
            Renderer = new SimpleRenderer(sls)
        };
        mapView1.GraphicsOverlays.Add(go);
        UpdateMapExtentGraphics(mapView1);
}
```

在构造函数中调用 InitMap 函数：

```
public MainWindow()
{
    InitializeComponent();
    InitMap();
}
```

地图视图更新后，重新设置下载窗口：

```
void UpdateMapExtentGraphics(MapView mapView)
{
    Viewpoint viewpoint = mapView.GetCurrentViewpoint(ViewpointType.BoundingGeometry);
    if (viewpoint == null)
        return;
    Envelope extent = viewpoint.TargetGeometry as Envelope;
    if (extent == null)
        return;
    EnvelopeBuilder eb = new EnvelopeBuilder(extent);
    eb.Expand(0.8);
    GraphicsOverlay go = mapView1.GraphicsOverlays.FirstOrDefault();
    if (go == null) return;
    Graphic graphic = go.Graphics.FirstOrDefault();
    if (graphic == null)
    {
        graphic = new Graphic(eb.ToGeometry());
        go.Graphics.Add(graphic);
    }
    else
    {
        graphic.Geometry = eb.ToGeometry();
```

        }
    }

用户点击下载 geodatabase 按钮后，开始下载：

```
private async void StartGeodatabaseGeneration()
{
    progressBar1.Visibility = Visibility.Visible;
    string gdbPath = Path.Combine(AppDomain.CurrentDomain.BaseDirectory, tbGdb.Text.Trim());
    //Create a task for generating a geodatabase.
    GeodatabaseSyncTask gdbSyncTask = await GeodatabaseSyncTask.CreateAsync(_featureServiceUri);
    //Get the graphic in the map view.
    Graphic redGraphic = mapView1.GraphicsOverlays.First().Graphics.First();
    //Get the current extent of the red preview box.
    Envelope extent = redGraphic.Geometry as Envelope;
    //Get the default parameters for the generate geodatabase task.
    GenerateGeodatabaseParameters generateParams = await gdbSyncTask.CreateDefaultGenerateGeodatabaseParametersAsync(extent);
    //Create a generate geodatabase job.
    GenerateGeodatabaseJob generateGdbJob = gdbSyncTask.GenerateGeodatabase(generateParams, gdbPath);
    //Handle the job changed event.
    generateGdbJob.JobChanged += GenerateGdbJobChanged;
    // Handle the progress changed event with an inline (lambda) function to show the progress bar.
    generateGdbJob.ProgressChanged += ((sender, e) =>
    {
        GenerateGeodatabaseJob job = sender as GenerateGeodatabaseJob;
        //Update the progress bar.
        UpdateProgressBar(job.Progress);
    });
    //Start the job.
    generateGdbJob.Start();
}
```

更新进度条：
```
private void UpdateProgressBar(int progress)
{
    Dispatcher.Invoke(() =>
    {
        progressBar1.Value = progress;
    });
}
```
处理 GenerateGeodatabaseJob 的状态事件：
```
private async void HandleGenerationStatusChange(GenerateGeodatabase
Job job)
{
    JobStatus status = job.Status;
    tbMsg.Text += $"GenerateGeodatabaseJob.Status={job.Status.
ToString()}" + Environment.NewLine;
    tbMsg.Text += $"\t{job.Messages.LastOrDefault().Message}" +
Environment.NewLine;
    // If the job completed successfully, add the geodatabase data to
the map.
    if (job.Status == JobStatus.Succeeded)
    {
        progressBar1.Visibility = Visibility.Collapsed;
        //Get the new geodatabase.
        Geodatabase resultGdb = await job.GetResultAsync();
        Process.Start(Path.GetDirectoryName(resultGdb.Path));
    }
    //See if the job failed.
    else if (job.Status == JobStatus.Failed)
    {
        //Show an error messageif failed.
        if (job.Error != null)
            tbMsg.Text += $"GenerateGeodatabaseJob.Error.Message=
{job.Error.Message}" + Environment.NewLine;
    }
}
```

【实验结果】

第 12 节　时间范围浏览器

时空大数据(spatial-temporal big data)包括时间、空间和专题属性信息,具有多源集成、海量存储、更新快速等特点。时态数据是时间维度信息,能够动态地播放,例如可以用颜色渐变结合柱状图动态地展示 GDP 变化,用动态的点展示台风中心移动的轨迹,这种动图比静态图具有更好的用户体验。

【实验目的】

通过设置感兴趣时间(time of interest,TOI),控制要素的显示。

【实验数据】

ArcGIS Online 在线地图服务:

https://sampleserver6.arcgisonline.com/arcgis/rest/services/Hurricanes/MapServer

【实验步骤】

1. 新建项目

.NET 框架:与 ArcGIS Runtime SDK 对应
模板列表:Visual C#/Windows/Windows Classic Desktop
模板:ArcGIS Runtime Application(WPF)
名称:TimeExtentViewer
位置:D:\ArcGISRuntimeTutorial

2. 编译项目

详情见第 1 章第 3 节。

3. 修改 MainWindow.xaml 标记语言

对地图视图,添加名称属性:

```xml
<esri:MapView Name="mapView1" Map="{Binding Map, Source={StaticResource MapViewModel}}" />
```
设计 UI：
```xml
<ToolBar VerticalAlignment="Top">
    <TextBlock Text="Year such 2000 or 2005" />
    <TextBox Name="tbYear" Text="2000" />
    <Button Name="btSetTimeExtent" Content="SetTimeExtent" Click="btSetTimeExtent_Click" />
    <Button Name="btClearTimeExtent" Content="ClearTimeExtent" Click="btClearTimeExtent_Click" />
</ToolBar>
```

4. 修改 MainWindow.xaml.cs 代码

添加命名空间：
```csharp
using Esri.ArcGISRuntime.Data;
using Esri.ArcGISRuntime.Mapping;
using System.IO;
```
生成全局变量，用于存储在线地图服务：
```csharp
string _hurricanes = "https://sampleserver6.arcgisonline.com/arcgis/rest/services/Hurricanes/MapServer";
```
编写 Initialize 函数：
```csharp
void Initialize()
{
    Uri _mapServerUri = new Uri(_hurricanes);
    //Load the layers from the corresponding URIs.
    ArcGISMapImageLayer imageryLayer = new ArcGISMapImageLayer(_mapServerUri);
    //FeatureLayer pointLayer = new FeatureLayer(_featureLayerUri);
    mapView1.Map.OperationalLayers.Add(imageryLayer);
    //mapView1.Map.OperationalLayers.Add(pointLayer);
}
```
在构造函数中调用 Initialize 函数：
```csharp
public MainWindow()
{
    InitializeComponent();
    Initialize();
}
void btSetTimeExtent_Click(object sender, RoutedEventArgs e)
```

```
}
    if (int.TryParse(tbYear.Text, out int year))
    {
        DateTime start = new DateTime(year, 1, 1);
        DateTime end = new DateTime(year, 12, 31);
        mapView1.TimeExtent = new TimeExtent(start, end);
    }
}
void btClearTimeExtent_Click(object sender, RoutedEventArgs e)
{
    mapView1.TimeExtent = null;
}
```

【实验结果】

范围设置前后对比图(上：未设置范围；左下：2000 年；右下：2005 年)

第 13 节 路径规划

路径规划是智能终端的重要功能。公众电子地图的主要功能包括兴趣点(线、面)查找、目标导航、业务订单和支付等功能。路径规划的理论基础是数据结构里的最短路径分析。

【实验目的】

设置起点和终点后，App 进行网络分析，计算规划路径，并给出行驶路线。本实验内容较多、综合性较强、逻辑复杂。

读者在熟悉本 App 的开发技术后，结合其他案例，如读取设备坐标、显示设备地址等内容，开发出一个综合的车载导航 App。

【实验数据】

ArcGIS Online 在线底图服务。

【实验步骤】

1. 新建项目

.NET 框架：与 ArcGIS Runtime SDK 对应

模板列表：Visual C#/Windows/Windows Classic Desktop

模板：ArcGIS Runtime Application（WPF）

名称：RouteSolver

位置：D：\ArcGISRuntimeTutorial

2. 编译项目

详情见第 1 章第 3 节。

3. 修改 MainWindow.xaml 标记语言

对地图视图，添加名称属性：

```
<esri:MapView Name = "mapView1" Map = "{Binding Map,Source = {StaticResource MapViewModel}}" />
```

设计 UI：

```
<StackPanel Opacity = "0.9" VerticalAlignment = "Top">
<ToolBar>
    <TextBlock Text = "Url" VerticalAlignment = "Center" />
    <TextBox Name = "tbUrl" MinWidth = "100" />
</ToolBar>
<ToolBar>
    <TextBlock Text = "from" />
    <TextBox Name = "tbFrom" MinWidth = "100" />
```

```xml
<TextBlock Text = "to" />
<TextBox Name = "tbTo" MinWidth = "100" />
<Button Content = "SetStops" Name = "btSetStops" Click = "btSetStops_Click" />
</ToolBar>
<Button Name = "SolveRouteButton" Content = "Solve Route" Click = "SolveRouteClick" HorizontalAlignment = "Left" />
<ListBox Width = "300" Name = "DirectionsListBox" HorizontalAlignment = "Left" ScrollViewer.HorizontalScrollBarVisibility = "Disabled" Height = "200"></ListBox>
</StackPanel>
```

4. 修改 MainWindow.xaml.cs 代码

添加命名空间:

```csharp
using Esri.ArcGISRuntime.Geometry;
using Esri.ArcGISRuntime.Mapping;
using Esri.ArcGISRuntime.Symbology;
using Esri.ArcGISRuntime.Tasks.NetworkAnalysis;
using Esri.ArcGISRuntime.UI;
```

生成全局变量，用于存储资源路径:

```csharp
private string _flag = @"ArcGISRuntimeSampleData\images\Symbols\Transportation\CheckeredFlag.png";
private string _car = @"ArcGISRuntimeSampleData\images\Symbols\Transportation\CarRedFront.png";
string _routeService = "http://sampleserver6.arcgisonline.com/arcgis/rest/services/NetworkAnalysis/SanDiego/NAServer/Route";
```

编写 Initialize 函数:

```csharp
void Initialize()
{
    tbUrl.Text = _routeService;
    //Define the route stop locations (points).
    MapPoint fromPoint = new MapPoint(-117.15494348793044, 32.706506537686927, SpatialReferences.Wgs84);
    MapPoint toPoint = new MapPoint(-117.14905088669816, 32.735308180609138, SpatialReferences.Wgs84);
    tbFrom.Text = fromPoint.ToText();
    tbTo.Text = toPoint.ToText();
    mapView1.GraphicsOverlays.Add(new GraphicsOverlay());
```

}

在构造函数中调用 Initialize 函数：
```
public MainWindow()
{
    InitializeComponent();
    Initialize ();
}
```

设置站点：
```
private void btSetStops_Click(object sender, RoutedEventArgs e)
{
    ClearStops();
    //Picture marker symbols: from = car, to = checkered flag.
    Uri carUri = new Uri(System.IO.Path.Combine(AppDomain.CurrentDomain.BaseDirectory + "..\\..\\..\\..\\", _car));
    Uri flagUri = new Uri(System.IO.Path.Combine(AppDomain.CurrentDomain.BaseDirectory + "..\\..\\..\\..\\", _flag));
    PictureMarkerSymbol carSymbol = new PictureMarkerSymbol(carUri);
    PictureMarkerSymbol flagSymbol = new PictureMarkerSymbol(flagUri);
    //Add a slight offset (pixels) to the picture symbols.
    carSymbol.OffsetX = -carSymbol.Width /2;
    carSymbol.OffsetY = -carSymbol.Height /2;
    flagSymbol.OffsetX = -flagSymbol.Width /2;
    flagSymbol.OffsetY = -flagSymbol.Height /2;
    //Create graphics for the stops.
    MapPoint fromPoint = MapPointEx.FromText(tbFrom.Text);
    MapPoint toPoint = MapPointEx.FromText(tbTo.Text);
    Graphic fromGraphic = new Graphic(fromPoint, carSymbol);
    Graphic toGraphic = new Graphic(toPoint, flagSymbol);
    //Create the graphics overlay and add the stop graphics.
    mapView1.GraphicsOverlays[0].Graphics.Add(fromGraphic);
    mapView1.GraphicsOverlays[0].Graphics.Add(toGraphic);
    //Get an Envelope that covers the area of the stops (and a little more).
    Envelope routeStopsExtent = new Envelope(fromPoint, toPoint);
    EnvelopeBuilder envBuilder =new EnvelopeBuilder(routeStopsExtent);
    envBuilder.Expand(1.5);
    //Create a new viewpoint apply it to the map view when the spatial reference changes.
```

```csharp
    Viewpoint sanDiegoViewpoint = new Viewpoint(envBuilder.ToGeometry());
    mapView1.SetViewpoint(sanDiegoViewpoint);
}
```

清除路径：

```csharp
private void ClearRouts()
{
    //Clear the list of directions.
    DirectionsListBox.ItemsSource = null;
    // Remove the route graphic from the graphics overlay (only line graphic in the collection).
    int graphicsCount = mapView1.GraphicsOverlays[0].Graphics.Count;
    for (int i = graphicsCount; i > 0; i--)
    {
        //Get this graphic and see if it has line geometry.
        Graphic g = mapView1.GraphicsOverlays[0].Graphics[i - 1];
        if (g.Geometry.GeometryType == GeometryType.Polyline)
        {
            //Remove the graphic from the overlay.
            mapView1.GraphicsOverlays[0].Graphics.Remove(g);
        }
    }
}
```

清除站点：

```csharp
private void ClearStops()
{
    // Remove the route graphic from the graphics overlay (only line graphic in the collection).
    int graphicsCount = mapView1.GraphicsOverlays[0].Graphics.Count;
    for (int i = graphicsCount; i > 0; i--)
    {
        //Get this graphic and see if it has line geometry.
        Graphic g = mapView1.GraphicsOverlays[0].Graphics[i - 1];
        if (g.Geometry.GeometryType == GeometryType.Point)
        {
            //Remove the graphic from the overlay.
            mapView1.GraphicsOverlays[0].Graphics.Remove(g);
        }
```

        }
    }
    路径规划：
async void SolveRouteClick(object sender, System.Windows.Routed EventArgs e)
{
    ClearRouts();
    //Create a new route task using the San Diego route service URI.
    //RouteTask solveRouteTask=await RouteTask.CreateAsync(_sanDiego RouteServiceUri);
    Uri serviceUri = new Uri(tbUrl.Text.Trim());
    RouteTask solveRouteTask=await RouteTask.CreateAsync(serviceUri);
    //Get the default parameters from the route task (defined with the service).
    RouteParameters routeParams=await solveRouteTask.CreateDefault ParametersAsync();
    //Make some changes to the default parameters.
    routeParams.ReturnStops = true;
    routeParams.ReturnDirections = true;
    //Set the list of route stops that were defined at startup.
    // Create Stop objects with the points and add them to a list of stops.
    MapPoint fromPoint = MapPointEx.FromText(tbFrom.Text);
    MapPoint toPoint = MapPointEx.FromText(tbTo.Text);
    Stop stop1 = new Stop(fromPoint);
    Stop stop2 = new Stop(toPoint);
    //_routeStops = new List<Stop> { stop1, stop2 };
    Stop[] stops = { stop1, stop2 };
    routeParams.SetStops(stops);
    // Solve for the best route between the stops and store the result.
    RouteResult solveRouteResult = await solveRouteTask.SolveRouteAsync(routeParams);
    //Get the first (should be only) route from the result.
    Route firstRoute = solveRouteResult.Routes.First();
    //Get the route geometry (polyline).
    Polyline routePolyline = firstRoute.RouteGeometry;
    //Create a thick purple line symbol for the route.

```
    SimpleLineSymbol routeSymbol = new SimpleLine
SymbolStyle.Solid, Color.Purple, 8.0);
    // Create a new graphic for the route geometry and add it to the
graphics overlay.
    Graphic routeGraphic = new Graphic(routePolyline, routeSymbol);
    mapView1.GraphicsOverlays[0].Graphics.Add(routeGraphic);
    //Get a list of directions for the route and display it in the list
box.
    IReadOnlyList<DirectionManeuver>directionsList=firstRoute.
DirectionManeuvers;
    DirectionsListBox.ItemsSource = directionsList;
    DirectionsListBox.DisplayMemberPath = "DirectionText";
}
```

【实验结果】

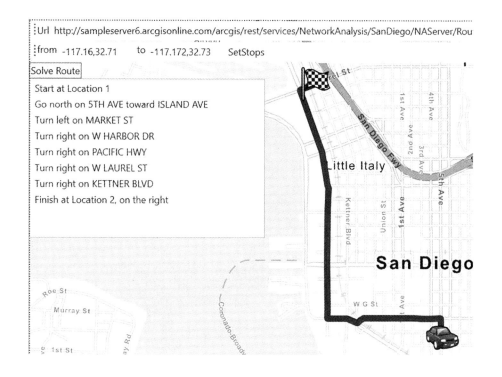

第 14 节　图层范围合并器

图层范围一般指的是所有要素范围的外接矩形区域，用东、南、西、北四个方位的坐标表示。

【实验目的】

获取每个图层的空间范围，并计算并集。

【实验数据】

ArcGIS Online 在线底图服务：

https：//sampleserver6.arcgisonline.com/arcgis/rest/services/World_Street_Map/MapServer

本地 shp 文件夹，包括江南五省的 5 个边界多边形：

ArcGISRuntimeSampleData \ Provinces

【实验步骤】

1. 新建项目

.NET 框架：与 ArcGIS Runtime SDK 对应

模板列表：Visual C#/Windows/Windows Classic Desktop

模板：ArcGIS Runtime Application（WPF）

名称：ExtentsCombiner

位置：D：\ArcGISRuntimeTutorial

2. 编译应用程序

详情见第 1 章第 3 节。

3. 修改 MainWindow.xaml 标记语言

对地图视图，添加名称属性：

```
<esri:MapView Name="mapView1" Map="{Binding Map,Source={StaticResource MapViewModel}}"/>
```

设计 UI：

```
<StackPanel Opacity="0.8">
    <TextBox Name="tbFile"/>
    <Button Name="btExtents" Content="Extents" Click="btExtents_Click"/>
    <TextBox Name="tbExtents"/>
</StackPanel>
```

4. 修改 MainWindow.xaml.cs 代码

添加命名空间：

```
using Esri.ArcGISRuntime.Data;
using Esri.ArcGISRuntime.Geometry;
using Esri.ArcGISRuntime.Mapping;
using System.IO;
```

生成全局变量,用于存储 shp 文件路径:
```
string _file = @"ChinaProvince\bou2_4p.shp";
```
编写 Initialize 函数:
```
void Initialize()
{
    string filepath = System.IO.Path.Combine(AppDomain.CurrentDomain.BaseDirectory + "..\\..\\..\\", _file);
    tbFile.Text = filepath;
}
```
在构造函数中调用 Initialize 函数:
```
public MainWindow()
{
    InitializeComponent();
    Initialize();
}
```
读取 shp 文件,获得每个图层的范围,最后进行合并:
```
async void btExtents_Click(object sender, RoutedEventArgs e)
{
    string filepath = tbFile.Text.Trim();
    string[] files = System.IO.Directory.GetFiles(filepath, "*.shp");
    List<Envelope> extents = new List<Envelope>();
    foreach (string file in files)
    {
        ShapefileFeatureTable sft = await ShapefileFeatureTable.OpenAsync(file);
        //Create a feature layer to display the shapefile.
        FeatureLayer fl = new FeatureLayer(sft);
        await fl.LoadAsync();
        mapView1.Map.OperationalLayers.Add(fl);
        //Add the extent to the list of extents.
        extents.Add(fl.FullExtent);
        tbExtents.Text += fl.Name + ":" + fl.FullExtent.ToString() + Environment.NewLine;
    }
    // Use the geometry engine to compute the full extent of the ENC Exchange Set.
    Envelope fullExtent = GeometryEngine.CombineExtents(extents);
```

```
            tbExtents.Text += "CombineExtents:" + fullExtent.ToString();
            //Set the viewpoint.
            mapView1.SetViewpoint(new Viewpoint(fullExtent));
}
```

【实验结果】

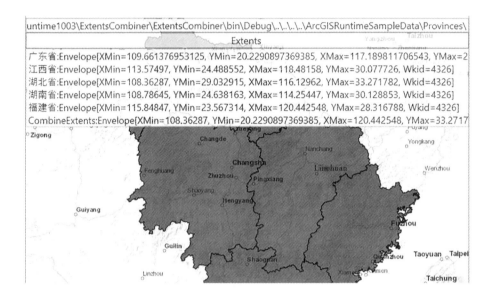

第7章 显示控制与渲染

地理信息可视化是运用图形学、计算机图形学和图像处理技术,将地学信息以图形、符号、图标、文字、表格、视频等形式显示并进行交互。

GIS 显示与渲染有其自身的特殊性,与 CAD 在显示方面有巨大的差异。CAD 一般是手绘(人工)控制显示样式,GIS 一般通过设置,自动化、批量化的进行显示控制。利用字段属性值进行显示设置,是 GIS 最常用的方法。

第 1 节 网格控制器

坐标网格(经纬网)可以直观显示当前位置,对目标进行快速定位,提高制图的专业性。

【实验目的】

交互式动态控制地图网格的显示或隐藏。

【实验数据】

ArcGIS Online 在线底图服务,需要网络连接。

【实验步骤】

1. 新建项目

.NET 框架:与 ArcGIS Runtime SDK 对应
模板列表:Visual C#/Windows/Windows Classic Desktop
模板:ArcGIS Runtime Application (WPF)
名称:GridController
位置:D:\ArcGISRuntimeTutorial

2. 编译项目

详情见第 1 章第 3 节

3. 修改 MainWindow.xaml 标记语言

对地图视图,添加名称属性:

```
<esri:MapView Name="mapView1" Map="{Binding Map,Source={StaticResource MapViewModel}}" />
```

设置 UI,添加一个层叠面板(StackPanel)容器,子控件包括文本框和按钮。

文本框，存储 shp 路径。

按钮，名称=btGird，用于控制网格的生成、显示和隐藏。

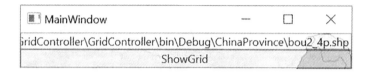

```
<StackPanel>
    <TextBox Name = "tbFile" Opacity = "0.8"/>
    <Button Name = "btGrid" Content = "ShowGrid" Click = "btGrid_Click" Opacity = "0.8"/>
</StackPanel>
```

4. 修改 MainWindow.xaml.cs 代码

添加命名空间：

```
using Esri.ArcGISRuntime.Data;
using Esri.ArcGISRuntime.Mapping;
using System.IO;
```

定义全局变量，存储 shp 文件位置：

```
string _shpPath = @"ChinaProvince\bou2_4p.shp";
```

编写 Initialize 函数：

```
private async void Initialize()
{
    //Get the path to the shapefile.
    string filepath = System.IO.Path.Combine(AppDomain.CurrentDomain.BaseDirectory, _shpPath);
    tbFile.Text = filepath;
    if (! File.Exists(filepath))
        return;
    //Open the shapefile.
    ShapefileFeatureTable myShapefile = await ShapefileFeatureTable.OpenAsync(filepath);
    //Create a feature layer to display the shapefile.
    FeatureLayer newFeatureLayer = new FeatureLayer(myShapefile);
    //Add the feature layer to the map.
    mapView1.Map.OperationalLayers.Add(newFeatureLayer);
    //Zoom the map to the extent of the shapefile.
```

```
    await mapView1.SetViewpointGeometryAsync(newFeatureLayer.
FullExtent);
}
```

在构造函数中调用 Initialize 函数：

```
public MainWindow()
{
    InitializeComponent();
    Initialize();
}
```

点击 ShowGrid 按钮事件：

```
private void btGrid_Click(object sender, RoutedEventArgs e)
{
    if (mapView1.Grid == null)
    {
        LatitudeLongitudeGrid grid = new LatitudeLongitudeGrid();
        mapView1.Grid = grid;
    }
    else
        mapView1.Grid.IsVisible = ! mapView1.Grid.IsVisible;
}
```

【实验结果】

点击 ShowGrid，显示网格；再点击此按钮，隐藏网格。

第 2 节　要素图层生成符号

Graphic 可以用图片（例如透明 PNG）进行标识，比要素类的原生显示方式具有更好的可视化效果和视觉冲击力。如果使用一些技巧，可以获得动画效果。

【实验目的】

根据本地 SHP 点文件获取地理坐标，通过 PNG 图片，模拟生成精美的机场分布图。

【实验数据】

本地 shp 文件，要素内容为小比例尺的中国各省级行政中心：
ArcGISRuntimeSampleData\ChinaProvince\res1_4m.shp

【实验步骤】

1. 新建项目

.NET 框架：与 ArcGIS Runtime SDK 对应

模板列表：Visual C#/Windows/Windows Classic Desktop

模板：ArcGIS Runtime Application（WPF）

名称：PictureMarkerSymbolFromFeatureLayer
位置：D：\ArcGISRuntimeTutorial

2. 编译项目

详情见第1章第3节。

3. 修改 MainWindow.xaml 标记语言

对地图视图，添加名称属性：

```xml
<esri:MapView Name="mapView1" Map="{Binding Map,Source={StaticResource MapViewModel}}"/>
```

设计 UI：

```xml
<StackPanel Opacity="0.8">
    <TextBox Name="tbShp"/>
    <TextBox Name="tbPic"/>
    <Button Content="LoadShp" Name="btLoadShp" Click="btLoadShp_Click"/>
</StackPanel>
```

4. 修改 MainWindow.xaml.cs 代码

添加命名空间：

```csharp
using Esri.ArcGISRuntime.Data;
using Esri.ArcGISRuntime.Mapping;
using System.IO;
```

生成全局变量，用于存储 shp 文件路径：

```csharp
string _file = @"ArcGISRuntimeSampleData\ChinaProvince\res1_4m.shp";
string _pic = @"ArcGISRuntimeSampleData\images\plane32East.png";
```

编写 Initialize 函数：

```csharp
void Initialize()
{
    string filepath = System.IO.Path.Combine(AppDomain.CurrentDomain.BaseDirectory + "..\\..\\..\\..\\", _file);
    string picpath = System.IO.Path.Combine(AppDomain.CurrentDomain.BaseDirectory + "..\\..\\..\\..\\", _pic);
    tbShp.Text = filepath;
    tbPic.Text = picpath;
}
```

在构造函数中调用 Initialize 函数：
```
public MainWindow()
{
    InitializeComponent();
    Initialize ();
}
```
添加 shp 文件，打开表格，生成符号(graphic)：
```
async void btLoadShp_Click(object sender, RoutedEventArgs e)
{
    string filepath = tbShp.Text;
    if (! File.Exists(filepath))
        return;
    string picpath = tbPic.Text;
    if (! File.Exists(picpath))
        return;
    ShapefileFeatureTable sft = await ShapefileFeatureTable.OpenAsync(filepath);
    mapView1.GraphicsOverlays.Add(new GraphicsOverlay());
    FileStream fs = new FileStream(picpath, FileMode.Open);
    PictureMarkerSymbol pms = await PictureMarkerSymbol.CreateAsync(fs);
    Graphic[] graphics = await GraphicCollectionEx.CreateFromFeatureTableAsync(sft, pms);
    mapView1.GraphicsOverlays[0].Graphics.AddRange(graphics);
    FeatureLayer fl = new FeatureLayer(sft);
    await fl.LoadAsync();
    mapView1.SetViewpoint(new Viewpoint(fl.FullExtent));
}
```
【实验结果】

第3节 栅格山体阴影渲染器

对数字高程模型进行山体阴影显示,可以获得三维立体效果。

【实验目的】

对栅格图层进行山体阴影渲染,可以根据太阳位置(方位角和高度角)调整灰度。

【实验数据】

存储在本地四川省阿坝藏族羌族自治州的数字高程模型:"dem \ 阿坝藏族羌族自治州.tif"。

【实验步骤】

1. 新建项目

.NET 框架:与 ArcGIS Runtime SDK 对应

模板列表:Visual C#/Windows/Windows Classic Desktop

模板:ArcGIS Runtime Application (WPF)

名称:RasterHillshadeRender

位置:D:\ArcGISRuntimeTutorial

2. 编译项目

详情见第1章第3节。

3. 修改 MainWindow.xaml 标记语言

给地图视图添加名称属性:

```
<esri:MapView Name="mapView1" Map="{Binding Map,Source={StaticResource MapViewModel}}"/>
```

添加网格控件和子控件,设置 UI 如下:

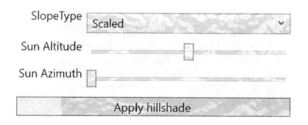

根据 UI,编写 xmal 语言:

```
<Grid Width="300" Background="White" Opacity="0.9" HorizontalAlignment="Left" VerticalAlignment="Top" Margin="10">
```

```xml
<Grid.ColumnDefinitions>
    <ColumnDefinition/>
    <ColumnDefinition Width="3*"/>
</Grid.ColumnDefinitions>
<Grid.RowDefinitions>
    <RowDefinition/>
    <RowDefinition/>
    <RowDefinition/>
    <RowDefinition/>
</Grid.RowDefinitions>
<TextBlock Grid.Row="0" Grid.Column="0"
    TextAlignment="Right"
    Text="SlopeType"/>
<ComboBox Name="cbSlopeType" Grid.Row="0" Grid.Column="1" Margin="5"/>

<TextBlock Grid.Row="1" Grid.Column="0"
    TextAlignment="Right"
    Text="Sun Altitude"/>
<Slider Name="sldAltitude" Grid.Row="1" Grid.Column="1" Minimum="0" Maximum="90" Margin="5">
    <Slider.ToolTip>
        <ToolTip Content="{Binding RelativeSource={RelativeSource Self},Path=PlacementTarget.Value}"
            ContentStringFormat="{}{0:0}" >
        </ToolTip>
    </Slider.ToolTip>
</Slider>
<TextBlock Grid.Row="2" Grid.Column="0"
    TextAlignment="Right"
    Text="Sun Azimuth"/>
<Slider Name="sldAzimuth" Grid.Row="2" Grid.Column="1" Minimum="0" Maximum="360" Margin="5">
    <Slider.ToolTip>
        <ToolTip Content="{Binding RelativeSource={RelativeSource Self},Path=PlacementTarget.Value}"
```

```
            ContentStringFormat = "{}{0:0}" />
        </Slider.ToolTip>
    </Slider>
    <Button Name = "btnApplyHillshade" Content = "Apply hillshade"
Click = "ApplyHillshadeButton_Click" Margin = "5" Grid.Row = "3"
Grid.Column
Span = "2" />
</Grid>
```

4. 修改 MainWindow.xaml.cs 代码

添加命名空间：

```
using Esri.ArcGISRuntime.Mapping;
using Esri.ArcGISRuntime.Rasters;
```

添加全局变量，用于存储山体阴影参数和栅格图层：

```
string _demPath = @"dem\阿坝藏族羌族自治州.tif";
//z 值缩放因子,用于单位转换或垂直夸张
double _zfactor = 1.0;
//像素尺寸权重和像素尺寸因子,用于适应视图缩放时的高度变化。
double _pixelSizePower = 1.0;
double _pixelSizeFactor = 1.0;
//像素比特深度
int _pixelBitDepth = 8;
//栅格图层
RasterLayer _rasterLayer;
```

编写 Initialize 函数，用于加载栅格数据，生成栅格图层，并添加到地图的业务图层中；同时设置 UI，填充坡度类型，供用户选择：

```
void Initialize()
{
    //Get the file name for the local raster dataset.
    String filepath = System.IO.Path.Combine(AppDomain.CurrentDomain.BaseDirectory, _demPath);
    //Load the raster file.
    Raster rasterFile = new Raster(filepath);
    //Create and load a new raster layer to show the image.
    _rasterLayer = new RasterLayer(rasterFile);
    await _rasterLayer.LoadAsync();
```

```
//Set the initial viewpoint with the raster's full extent.
mapView1.Map.InitialViewpoint = new Viewpoint(_rasterLayer.FullExtent);
//Add the layer to the map.
mapView1.Map.OperationalLayers.Add(_rasterLayer);
//Add slope type values to the combo box.
foreach (SlopeType slope in Enum.GetValues(typeof(SlopeType)))
{
    cbSlopeType.Items.Add(slope);
}
//Select the "Scaled" slope type enum.
cbSlopeType.SelectedValue = SlopeType.Scaled;
}
```

根据参数生成山体阴影：

```
private void ApplyHillshadeButton_Click(object sender, System.Windows.RoutedEventArgs e)
{
    double altitude = sldAltitude.Value;
    double azimuth = sldAzimuth.Value;
    SlopeType st = (SlopeType)cbSlopeType.SelectedItem;
    //Create a hillshade renderer that uses the values selected by the user.
    HillshadeRenderer hillshadeRenderer = new HillshadeRenderer(altitude, azimuth, _zfactor, st, _pixelSizeFactor, _pixelSizePower, _pixelBitDepth);
    //Apply the new renderer to the raster layer.
    _rasterLayer.Renderer = hillshadeRenderer;
}
```

在构造函数中调用 Initialize 函数：

```
public MainWindow()
{
    InitializeComponent();
    Initialize ();
}
```

【实验结果】

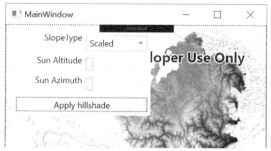 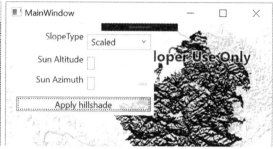

第4节 电子海图浏览器

电子海图(electronic navigational charts，ENCs)是对水文和海洋信息进行空间可视化和分析的矢量数据集。Runtime 支持国际水文标准(International Hydrographical Organization，IHO)S-57 标准。

【实验目的】

以专业的标准样式，显示电子海图。

【实验数据】

ArcGIS Online 在线底图服务：

https：//sampleserver6.arcgisonline.com/arcgis/rest/services/World_Street_Map/MapServer

本地电子海图 ENC 文件：

`ArcGISRuntimeSampleData\Enc_ExchangeSetwithoutUpdates\ExchangeSetwithoutUpdates\ENC_ROOT\CATALOG.031`

【实验步骤】

1. 新建项目

.NET 框架：与 ArcGIS Runtime SDK 对应

模板列表：Visual C#/Windows/Windows Classic Desktop

模板：ArcGIS Runtime Application (WPF)

名称：EncViewer

位置：D：\ArcGISRuntimeTutorial

2. 编译应用程序

详情见第1章第3节。

3. 修改 MainWindow.xaml 标记语言

对地图视图，添加名称属性：
<esri:MapView Name="mapView1" Map="{Binding Map,Source={Static Resource MapViewModel}}" />

4. 添加 Esri.ArcGISRuntime.Hydrography NuGet 包：

添加步骤如下：

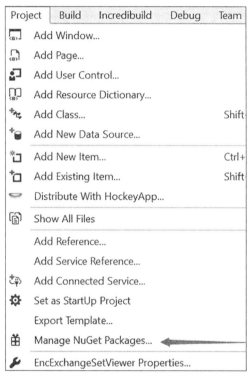

5. 修改 MainWindow.xaml.cs 代码

添加命名空间：

```
using Esri.ArcGISRuntime.Data;
using Esri.ArcGISRuntime.Mapping;
using Esri.ArcGISRuntime.Hydrography;
using System.IO;
```

生成全局变量，用于存储 shp 文件路径：

```
string _file = @"ArcGISRuntimeSampleData\Enc_ExchangeSetwithout Updates\ExchangeSetwithoutUpdates\ENC_ROOT\CATALOG.031";
```

编写 Initialize 函数：

```
void Initialize()
{
    //Get the path to the enc file
    string filepath = System.IO.Path.Combine(AppDomain.CurrentDomain.BaseDirectory + "..\\..\\..\\..\\", _file);
    tbFile.Text = filepath;
}
```

在构造函数中调用 Initialize 函数：

```
public MainWindow()
{
    InitializeComponent();
    Initialize();
}
```

点击按钮，添加海图：

```
async void btOpen_Click(object sender, RoutedEventArgs e)
{
    string filepath = tbFile.Text.Trim();
    //Create the Exchange Set.
    // Note: this constructor takes an array of paths because so that update sets can be loaded alongside base data.
    EncExchangeSet myEncExchangeSet = new EncExchangeSet(filepath);
    //Wait for the exchange set to load.
    await myEncExchangeSet.LoadAsync();
    // Store a list of data set extent's - will be used to zoom the mapview to the full extent of the Exchange Set.
    List<Envelope> dataSetExtents = new List<Envelope>();
    //Add each data set as a layer.
```

```
foreach (EncDataset myEncDataset in myEncExchangeSet.Datasets)
{
    //Create the cell and layer.
    EncLayer myEncLayer = new EncLayer(new EncCell(myEncDataset));
    //Add the layer to the map.
    mapView1.Map.OperationalLayers.Add(myEncLayer);
    //Wait for the layer to load.
    await myEncLayer.LoadAsync();
    //Add the extent to the list of extents.
    dataSetExtents.Add(myEncLayer.FullExtent);
}
// Use the geometry engine to compute the full extent of the ENC Exchange Set.
Envelope fullExtent = GeometryEngine.CombineExtents(dataSetExtents);
//Set the viewpoint.
mapView1.SetViewpoint(new Viewpoint(fullExtent));
}
```

【实验结果】

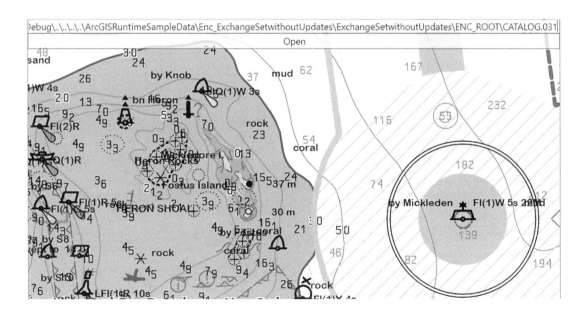

第5节 军事符号地图

各行业都有自己的地图标准格式或规范。海图、军图是常见的行业标准地图案例。林业、水利、国土、航空、交通也有自己的行业制图规范。

【实验目的】

将字典渲染器应用于要素图层并显示 MIL25255D 图形。字典渲染器使用 mil2525d 样式文件和要素的属性创建军事符号图形。

【实验数据】

ArcGIS Online 在线底图服务：

https：//sampleserver6.arcgisonline.com/arcgis/rest/services/World_Street_Map/MapServer

本地地理数据库：

ArcGISRuntimeSampleData \ militaryoverlay_geodatabase \ militaryoverlay.geodatabase

本地军事样式符号库：

ArcGISRuntimeSampleData \ mil2525d_stylx \ mil2525d.stylx

【实验步骤】

1. 新建项目

.NET 框架：与 ArcGIS Runtime SDK 对应

模板列表：Visual C#/Windows/Windows Classic Desktop

模板：ArcGIS Runtime Application（WPF）

名称：MilitarySymbology

位置：D：\ArcGISRuntimeTutorial

2. 编译项目

详情见第1章第3节。

3. 修改 MainWindow.xaml 标记语言

对地图视图，添加名称属性：

```
<esri:MapView Name="mapView1" Map="{Binding Map,Source={Static
Resource MapViewModel}}" />
```

设计 UI，添加一个层叠面板，子控件包括2个文本框和2个按钮。

文本框：用于输入地理数据库(.geodatabase)路径。

文本框：用于输入军事样式库(.stylx)路径。

按钮：切换(显示或者隐藏)普通要素符号图层。

按钮：显示军用地图。

代码如下：

```xml
<StackPanel>
    <TextBox Name="tbGdb"/>
    <TextBox Name="tbStylx"/>
    <Button Name="btFeature" Content="Show/Hide FeatureSymbology" Click="btFeature_Click"/>
    <Button Name="btMilitary" Content="MilitarySymbology" Click="btMilitary_Click"/>
</StackPanel>
```

4. 修改 MainWindow.xaml.cs 代码

添加命名空间：

```csharp
using Esri.ArcGISRuntime.Data;
using Esri.ArcGISRuntime.Geometry;
using Esri.ArcGISRuntime.Mapping;
using Esri.ArcGISRuntime.Symbology;
using System.IO;
```

生成全局变量，用于存储军事样式库和地理数据库路径：

```csharp
string _styleFile = @"ArcGISRuntimeSampleData\mil2525d_stylx\mil2525d.stylx";
string _gdbFile = @"ArcGISRuntimeSampleData\militaryoverlay_geodatabase\militaryoverlay.geodatabase";
```

编写 Initialize 函数：

```csharp
void Initialize()
{
    //Create geometry for the center of the map.
    MapPoint center = new MapPoint(-13549402.587055, 4397264.96879385, SpatialReference.Create(3857));
    //Set the map's viewpoint to highlight the desired content.
    await mapView1.SetViewpointAsync(new Viewpoint(center, 201555));
    tbStylx.Text = System.IO.Path.Combine(AppDomain.CurrentDomain.BaseDirectory + "..\\..\\..\\..\\", _styleFile);
    tbGdb.Text = System.IO.Path.Combine(AppDomain.CurrentDomain.BaseDirectory + "..\\..\\..\\..\\", _gdbFile);
}
```

在构造函数中调用 Initialize 函数：

```csharp
public MainWindow()
{
    InitializeComponent();
```

```
        Initialize ();
}
    显示/隐藏普通地图：
async void btFeature_Click(object sender, RoutedEventArgs e)
{
    if (mapView1.Map.OperationalLayers.Count == 0)
    {
        //Get the path to the geodatabase.
        string geodbFilePath = tbGdb.Text.Trim();
        if (! File.Exists(geodbFilePath))
            return;
        //Load the geodatabase from local storage.
        Geodatabase baseGeodatabase = await Geodatabase.OpenAsync(geodbFilePath);
        // Add geodatabase features to the map, using the defined symbology.
        foreach (FeatureTable table in baseGeodatabase.GeodatabaseFeatureTables)
        {
            //Load the table.
            await table.LoadAsync();
            //Create the feature layer from the table.
            FeatureLayer fl = new FeatureLayer(table);
            //Load the layer.
            await fl.LoadAsync();
            //Add the layer to the map.
            mapView1.Map.OperationalLayers.Add(fl);
        }
    }
    else
    {
        foreach (FeatureLayer fl in mapView1.Map.OperationalLayers)
        {
            fl.IsVisible = ! fl.IsVisible;
        }
    }
}
    显示军用地图：
async void btMilitary_Click(object sender, RoutedEventArgs e)
```

```csharp
{
    //Get the path to the geodatabase.
    string geodbFilePath = tbGdb.Text.Trim();
    if (! File.Exists(geodbFilePath))
        return;
    //Load the geodatabase from local storage.
    Geodatabase baseGeodatabase = await Geodatabase.OpenAsync(geodbFilePath);
    //Get the path to the symbol dictionary.
    string symbolFilepath = tbStylx.Text;
    if (! File.Exists(symbolFilepath))
        return;
    //Load the symbol dictionary from local storage.
    // Note that the type of the symbol definition must be explicitlyprovided along with the file name.
    DictionarySymbolStyle dss = await DictionarySymbolStyle.OpenAsync("mil2525d", symbolFilepath);
    //Add geodatabase features to the map, using the defined symbology.
    foreach (FeatureTable table in baseGeodatabase.GeodatabaseFeatureTables)
    {
        //Load the table.
        await table.LoadAsync();
        //Create the feature layer from the table.
        FeatureLayer fl = new FeatureLayer(table);
        //Load the layer.
        await fl.LoadAsync();
        //Create a Dictionary Renderer using the DictionarySymbolStyle.
        DictionaryRenderer dictRenderer = new DictionaryRenderer(dss);
        //Apply the dictionary renderer to the layer.
        fl.Renderer = dictRenderer;
        //Add the layer to the map.
        mapView1.Map.OperationalLayers.Add(fl);
    }
}
```

【实验结果】

使用方法：

(1)在军事样式文本框中输入样式文件路径(.stylx)；

(2)在地理数据库文本框输入地理数据库路径(.geodatabase)；

(3)点击显示/隐藏要素符号(Show/Hide FeatureSymbology)，以默认样式显示普通要素地图(下图左)；

(4)点击显示/隐藏要素符号(Show/Hide FeatureSymbology)，隐藏普通要素地图；

(5)点击军事符号(MilitarySymbology)，生成军用地图(下图右)；

(6)对比普通地图和军用地图，可以发现，两幅地图的符号主要差别是军事地图中，箭头符号可以更加清晰地表达方向信息，同时点状符号也更符合军用制图习惯。

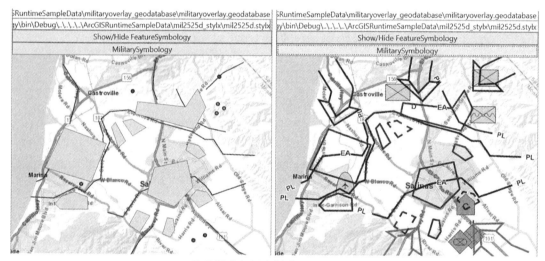

普通样式与军事样式显示对比图

第6节 视点管理器

视点(View point)设置地图显示中心、范围(比例尺)和旋转角度，在导航中可以根据设备坐标和前进方向，动态更新地图。

【实验目的】

通过设置视点坐标、比例尺和旋转角度，换个方式看地图。熟悉以下 API 的使用：

SetViewpointCenterAsync

SetViewpointScaleAsync

SetViewpointRotationAsync

【实验数据】

北部湾大学地理坐标：经度 = 108.61，纬度 = 21.96。

比例尺 = 1 : 1 000 000

旋转角度=90

【实验步骤】

1. 新建项目

.NET 框架：与 ArcGIS Runtime SDK 对应
模板列表：Visual C#/Windows/Windows Classic Desktop
模板：ArcGIS Runtime Application（WPF）
名称：ViewpointManager
位置：D：\ArcGISRuntimeTutorial

2. 编译项目

详情见第 1 章第 3 节。

3. 修改 MainWindow.xaml 标记语言

```
<Grid.RowDefinitions>
    <RowDefinition Height="auto"/>
    <RowDefinition Height="*"/>
</Grid.RowDefinitions>
<ToolBar Grid.Row="0">
    <TextBox Text="108.61" Name="tbLongitue"/>
    <TextBox Text="21.96" Name="tbLatitude"/>
    <Button Content="Center" Name="btCenter" Click="btCenter_Click"/>
    <Separator/>
    <TextBox Text="1000000" Name="tbScale"/>
    <Button Content="Scale" Name="btScale" Click="btScale_Click"/>
    <Separator/>
    <TextBox Text="90" Name="tbRotate"/>
    <Button Content="Rotate" Name="btRotate" Click="btRotate_Click"/>
</ToolBar>
<esri:MapView Name="mapView1" Grid.Row="1" Map="{Binding Map, Source={StaticResource MapViewModel}}"/>
```

4. 修改 MainWindow.xaml.cs 代码

```
private async void btCenter_Click(object sender, RoutedEventArgs e)
{
    double.TryParse(tbLongitue.Text, out double longitude);
```

```
    double.TryParse(tbLatitude.Text, out double latitude);
    await mapView1.SetViewpointCenterAsync(longitude, latitude);
}
private async void btScale_Click(object sender, RoutedEventArgs e)
{
    double.TryParse(tbScale.Text, out double scale);
    await mapView1.SetViewpointScaleAsync(scale);
}
private async void btRotate_Click(object sender, RoutedEventArgs e)
{
    double.TryParse(tbRotate.Text, out double rotate);
    await mapView1.SetViewpointRotationAsync(rotate);
}
```

【实验结果】

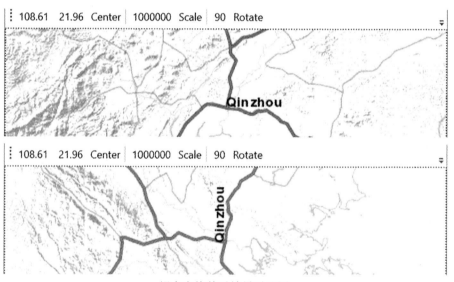

视点变换前后效果对比图

第 7 节　模拟飞行三维动画

在 GIS 三维场景中进行飞行模拟，具有数据真实、场景真实、细节真实的优点，比虚拟三维场景具有更高的决策效果。

【实验目的】

在夏威夷地区模拟三维地形飞行。本节实验较为复杂，需要反复研读体会。

【实验数据】

ArcGIS 在线高程服务：

http：//elevation3d.arcgis.com/arcgis/rest/services/WorldElevation3D/Terrain3D/ImageServer

【实验步骤】

1. 新建项目

.NET 框架：与 ArcGIS Runtime SDK 对应
模板列表：Visual C#/Windows/Windows Classic Desktop
模板：ArcGIS Runtime Application（WPF）
名称：Animate3D
位置：D：\ArcGISRuntimeTutorial

2. 编译项目

详情见第 1 章第 3 节。
对场景视图，添加名称属性。

3. 修改 MainWindow.xaml 标记语言

```
<esri:SceneView Name="sceneView1"/>
```

对地图视图，添加名称，设置大小(200*200)和位置(左下角)：

```
<esri:MapView Width="200" Height="200" Opacity="0.6" VerticalAlignment="Bottom" HorizontalAlignment="Left" Name="mapView1"/>
```

在右上角添加文本框，设置大小(100*100)，背景白色，透明显示：

```
<TextBox Name="tbInfo" Width="100" Height="100" Background="White" Opacity="0.6" HorizontalAlignment="Right" VerticalAlignment="Top"/>
```

4. 修改 MainWindow.xaml.cs 代码

添加 System.Drawing 程序集引用：
添加命名空间：

```
using Esri.ArcGISRuntime.Geometry;
using Esri.ArcGISRuntime.Mapping;
using Esri.ArcGISRuntime.Symbology;
using Esri.ArcGISRuntime.UI;
using System.Drawing;
using System.IO;
using System.Timers;
```

生成全局变量，存储飞机三维模型路径：

```csharp
string _plan3D = @"ArcGISRuntimeSampleData\Bristol_dae\Bristol.
dae";
```
生成全局变量，存储飞机飞行路径坐标：
```csharp
string _hawaii = @"ArcGISRuntimeSampleData\Animate3D_csv\Hawaii.
csv";//夏威夷
Graphic _gPlane3D; //飞机3D模型。
Graphic _gPlane2D; //平面地图上飞机位置,用三角形表示
Graphic _gRoute; //飞机飞行路径。
int _keyFrame; //飞机当前飞行场景帧索引。
int _frameCount; //飞机飞行总帧数。
(double Longitude, double Latitude, double Elevation, double Heading,
double Pitch, double Roll)[] _frames; //飞机飞行帧坐标集合。
```
编写 Init 函数：
```csharp
async void Init()
{
    sceneView1.Scene = new Scene(Basemap.CreateImagery());
    mapView1.Map = new Map(Basemap.CreateImagery());
    Uri elevUri = new Uri(_elevation);
    ElevationSource elevationSource = new ArcGISTiledElevationSource
(elevUri);
    Surface surface = new Surface();
    surface.ElevationSources.Add(elevationSource);
    sceneView1.Scene.BaseSurface = surface;
    GraphicsOverlay goPlane3D = new GraphicsOverlay();
    goPlane3D.SceneProperties.SurfacePlacement = SurfacePlacement.
Absolute;
    SimpleRenderer sr3D = new SimpleRenderer();
    sr3D.SceneProperties.HeadingExpression="[HEADING]";
    sr3D.SceneProperties.PitchExpression="[PITCH]";
    sr3D.SceneProperties.RollExpression="[ROLL]";
    goPlane3D.Renderer = sr3D;
    sceneView1.GraphicsOverlays.Add(goPlane3D);
    string planePath = System.IO.Path.Combine(AppDomain.Current
Domain.BaseDirectory + "..\\..\\..\\..\\", _plan3D);
    ModelSceneSymbol mss = await ModelSceneSymbol.CreateAsync( new
Uri(planePath), 1);
    _gPlane3D = new Graphic(new MapPoint(0, 0, SpatialReferences.
Wgs84), mss);
```

```
    goPlane3D.Graphics.Add(_gPlane3D);
    SimpleMarkerSymbol sms = new SimpleMarkerSymbol(SimpleMarker
SymbolStyle.Triangle, Color.Blue, 10);
    SimpleRenderer sr2D = new SimpleRenderer(sms);
    sr2D.RotationExpression="[ANGLE]";
    GraphicsOverlay goPlane2D = new GraphicsOverlay() { Renderer =
sr2D };
    mapView1.GraphicsOverlays.Add(goPlane2D);
    Dictionary<string, object> plane2DAttributes = new Dictionary<
string, object>();
    plane2DAttributes["ANGLE"] = 0;
    _gPlane2D = new Graphic(new MapPoint(0, 0, SpatialReferences.
Wgs84), plane2DAttributes);
    goPlane2D.Graphics.Add(_gPlane2D);
    GraphicsOverlay goRoute = new GraphicsOverlay();
    SimpleLineSymbol sls = new SimpleLineSymbol(SimpleLineSymbol
Style.Solid, Color.Red, 2);
    _gRoute = new Graphic() { Symbol = sls };
    goRoute.Graphics.Add(_gRoute);
    mapView1.GraphicsOverlays.Add(goRoute);
    OrbitGeoElementCameraController ogecc = new OrbitGeoElement
CameraController(_gPlane3D, 20);
    ogecc.CameraPitchOffset = 75;
    sceneView1.CameraController = ogecc;
    _keyFrame = 0;
    _frames = GetFrames();
    _frameCount = _frames.Length;
    //Draw mission route on the inset.
    //Create a collection of points to hold the mission.
    PointCollection points = new PointCollection(SpatialReferences.
Wgs84);
    //Add all of the points from the mission to the point collection.
    points.AddPoints(_frames.Select(m => new MapPoint(m.Longitude,
m.Latitude, m.Elevation, SpatialReferences.Wgs84)));
    // Create a polyline to symbolize the route from the point
collection.
    Polyline route = new Polyline(points);
    //Update the route graphic's geometry with the newly created route
```

```
polyline.
    _gRoute.Geometry = route;
    //Update the inset map's scale.
    await mapView1.SetViewpointScaleAsync(100000);
    Timer timer = new Timer(60)
    {
        Enabled = true,
        AutoReset = true
    };
    timer.Elapsed += Timer_Elapsed;
}
```

在构造函数中调用 Init 函数：

```
public MainWindow()
{
    InitializeComponent();
    Init ();
}
```

获取飞机飞行帧：

```
(double Longitude, double Latitude, double Elevation, double Heading, double Pitch, double Roll)[] GetFrames()
{
    //Get the path to the file.
    string filePath = System.IO.Path.Combine(AppDomain.CurrentDomain.BaseDirectory + "..\\..\\..\\..\\", _Hawaii);
    //Read the file text.
    var fileContents = File.ReadLines(filePath);
    //Split the file contents into a list of lines.
    var result = fileContents
    //remove any null MissionFrames.
    .Where(line => line != null)
    //for each line, create a MissionFrame object.
    .Select(line =>
    {
        string[] data = line.Split(',');
        return (Convert.ToDouble(data[0]), Convert.ToDouble(data[1]), Convert.ToDouble(data[2]), Convert.ToDouble(data[3]), Convert.ToDouble(data[4]), Convert.ToDouble(data[5]));
    }).ToArray();// Finally return that list of MissionFrames as an
```

array.
    return result;
}
   Timer 激发事件，实现飞机飞行：
void Timer_Elapsed(object sender, ElapsedEventArgs e)
{
    //Get the next position; % prevents going out of bounds even if the keyframe value is
    // changed unexpectedly (e.g. due to user interaction with the progress slider).
     (double Longitude, double Latitude, double Elevation, double Heading, double Pitch, double Roll) currentFrame = _frames[_keyFrame % _frameCount];
    // This is needed because the event could be running on a non-UI thread
    Dispatcher.BeginInvoke(new Action(() =>
    { //Update stats display
        tbInfo.Text = $"Elevation={currentFrame.Elevation}\n Heading={currentFrame.Heading}\n Pitch={currentFrame.Pitch}\n Roll={currentFrame.Roll}\n";
    }));
    //Update plane's position.
     MapPoint mpCurrent = new MapPoint(currentFrame.Longitude, currentFrame.Latitude, currentFrame.Elevation, SpatialReferences.Wgs84);
    _gPlane3D.Geometry = mpCurrent;
    _gPlane3D.Attributes["HEADING"] = currentFrame.Heading;
    _gPlane3D.Attributes["PITCH"] = currentFrame.Pitch;
    _gPlane3D.Attributes["ROLL"] = currentFrame.Roll;
    //Update the inset map; plane symbol position.
    _gPlane2D.Geometry = mpCurrent;
    //Update inset's viewpoint and heading.
    Viewpoint vp = new Viewpoint(mpCurrent, mapView1.MapScale, 360 + (float)currentFrame.Heading);
    mapView1.SetViewpoint(vp);
    //Update the keyframe. This advances the animation.
    _keyFrame++;
    //Restart the animation if it has finished.

```
    if (_keyFrame >= _frameCount)
    {
        _keyFrame = 0;
    }
}
```

【实验结果】

第 8 章　地理信息处理服务

地理信息服务(geographic information service)可以提高地理信息系统的利用率,建立一种面向服务的商业模式,用户可以按需获得和使用地理数据和计算服务。

简单的地理处理服务,相当于工具,在本地完成计算。大型的地理处理服务借助于万维网来展示,在 Web 应用程序中收集数据,以要素、地图、报表及文件等新式返回客户端。

复杂的地理处理任务是一个运行在服务器上的地理处理工具,执行和输出是通过服务器管理的,本章最后的案例就是通过本地服务器运行的。

第 1 节　基于位置的三维视线分析

基于位置的三维视线分析(可视性分析),是运用计算几何原理和技术,解决地形上控制点集合的可视范围。

可视分析的类型:
(1)点对的通视区域计算;
(2)点对的通视性判断;
(3)多点通视范围的交集;
(4)由可视范围反求待定位置与高度。

在三维场景下,可视分析可应用于旅游地理风景评价、房地产视线遮挡判断、通信信号覆盖、军事火力覆盖等许多场景,具有计算结果准确、展示直观等优点。

【实验目的】

选择珠穆朗玛峰附近区域进行三维视线分析。在地图上第一次点击,选取观察点;第二次点击,选取目标点,生成视线的可视范围,用不同的颜色表示可见部分和遮挡部分。

【实验数据】

ArcGIS 在线三维高程服务:

http://elevation3d.arcgis.com/arcgis/rest/services/WorldElevation3D/Terrain3D/ImageServer

珠穆朗玛峰附近坐标:

MapPoint(87, 28, SpatialReferences.Wgs84)

【实验步骤】

1. 新建项目

.NET 框架：与 ArcGIS Runtime SDK 对应
模板列表：Visual C#/Windows/Windows Classic Desktop
模板：ArcGIS Runtime Application (WPF)
名称：LocationBasedLineOfSight
位置：D：\ArcGISRuntimeTutorial

2. 编译项目

详情见第 1 章第 3 节。

3. 修改 MainWindow.xaml 标记语言

对地图视图，添加名称属性：
<esri：MapView Name = " mapView1" Map = " {Binding Map, Source = {StaticResource MapViewModel}}"/>

设计 UI，添加 2 个文本框 tbObserver、tbTarget，用于显示观察点和目标点的坐标。

4. 修改 MainWindow.xaml.cs 代码

添加命名空间：
```
using Esri.ArcGISRuntime.Geometry;
using Esri.ArcGISRuntime.Mapping;
using Esri.ArcGISRuntime.UI;
using Esri.ArcGISRuntime.UI.GeoAnalysis;
```
生成全局变量，用于存储 shp 文件路径：
```
//URL for an image service to use as an elevation source.
string _elevation = @"http://elevation3d.arcgis.com/arcgis/rest/services/WorldElevation3D/Terrain3D/ImageServer";
//Location line of sight analysis.
LocationLineOfSight _lineOfSight;
//Observer location for line of sight.
MapPoint _observer;
//Target location for line of sight.
MapPoint _target;
```
编写 Init 函数：
```
void Init()
{
    //Create a new Scene with an imagery basemap.
    Scene scene = new Scene(Basemap.CreateImagery());
```

```csharp
//Create an elevation source for the Scene.
ArcGISTiledElevationSource elevationSrc=new ArcGISTiled
ElevationSource(new Uri(_elevation));
scene.BaseSurface.ElevationSources.Add(elevationSrc);
//Add the Scene to the SceneView.
sceneView1.Scene = scene;
//Set the viewpoint with a new camera.
Camera camera = new Camera(new MapPoint(87, 28, Spatial
References.Wgs84), 10000, 0, 45, 0);
sceneView1.SetViewpointCameraAsync(camera);
// Create a new line of sight analysis with arbitrary points
(observer and target will be defined by the user).
_lineOfSight = new LocationLineOfSight ( new MapPoint ( 0, 0,
SpatialReferences.Wgs84), new MapPoint(0, 0, SpatialReferences.
Wgs84));
LineOfSight.LineWidth *= 3;
//Create an analysis overlay to contain the analysis and add it to
the scene view.
AnalysisOverlay ao = new AnalysisOverlay();
ao.Analyses.Add(_lineOfSight);
sceneView1.AnalysisOverlays.Add(ao);
}
```

在构造函数中调用 Init 函数：

```csharp
public MainWindow()
{
    InitializeComponent();
    Init ();
}
```

场景视图点击事件：

```csharp
void sceneView1_GeoViewTapped(object sender, Esri.ArcGISRuntime.
UI.Controls.GeoViewInputEventArgs e)
{
    //Ignore if tapped out of bounds (e.g. the sky).
    if (e.Location == null)
    {
        return;
    }
    //When the view is tapped, define the observer or target location
    with the tap point as appropriate.
```

```
    if (_observer == null)
    {
        // Define the observer location and set the target to the same point.
        _observer = new MapPoint(e.Location.X, e.Location.Y, e.Location.Z);
        _lineOfSight.ObserverLocation = _observer;
        _lineOfSight.TargetLocation = _observer;
        tbObserver.Text = _observer.ToString();
        // Clear the target location (if any) so the next click will define the target.
        _target = null;
    }
    else if (_target == null)
    {
        //Define the target.
        _target = new MapPoint(e.Location.X, e.Location.Y, e.Location.Z);
        _lineOfSight.TargetLocation = _target;
        tbTarget.Text = _target.ToString();
        //Clear the observer location so it can be defined again
        _observer = null;
    }
}
```

【实验结果】

导航地图到合适的位置和视角后，软件自动生成视线的可视范围，其中绿色表示可见部分，红色表示遮挡部分。

第 2 节　基于图形的三维视线分析

将瞭望台设置为监测点，对运动车辆进行监测，以运动车辆作为巡逻车，对周围环境进行巡视，是常见的三维视域分析任务。

【实验目的】

以高程影像和三维建筑组成三维场景，点击鼠标设置观察点，将运动车辆设置为检测目标。App 实时计算从观察点到目标点的通视性。可视部分和遮挡部分用不同颜色区分。

【实验数据】

ArcGIS 在线三维高程服务：

http：//elevation3d.arcgis.com/arcgis/rest/services/WorldElevation3D/Terrain3D/ImageServer

ArcGIS 在线建筑三维场景服务：

https://tiles.arcgis.com/tiles/z2tnIkrLQ2BRzr6P/arcgis/rest/services/New_York_LoD2_3D_Buildings/SceneServer/layers/0

本地汽车三维模型：ArcGISRuntimeSampleData\dolmus_3ds\dolmus.3ds

【实验步骤】

1. 新建项目

.NET 框架：与 ArcGIS Runtime SDK 对应

模板列表：Visual C#/Windows/Windows Classic Desktop

模板：ArcGIS Runtime Application（WPF）

名称：LineOfSighter

位置：D：\ArcGISRuntimeTutorial

2. 编译项目

详情见第 1 章第 3 节。

3. 修改 MainWindow.xaml 标记语言

删除地图视图，添加场景视图，添加名称属性：

```
<esri:SceneView Name = "sceneView1" GeoViewTapped = "sceneView1_GeoViewTapped"/>
```

设计 UI，加入状态栏，添加文本框，用于展示视线通视(可见/遮挡)。

```
<StatusBar Name="statusBar1" VerticalAlignment="Bottom">
    <TextBox Name="tbLineOfSight" Margin="3"/>
</StatusBar>
```

4. 修改 MainWindow.xaml.cs 代码

添加 System.Drawing 引用。

添加命名空间：

```csharp
using Esri.ArcGISRuntime.Geometry;
using Esri.ArcGISRuntime.Mapping;
using Esri.ArcGISRuntime.Symbology;
using Esri.ArcGISRuntime.UI;
using Esri.ArcGISRuntime.UI.GeoAnalysis;
using System;
using System.Timers;
using System.Drawing;
using System.Windows;
using System.Windows.Threading;
using Esri.ArcGISRuntime.UI.Controls;
```

生成全局变量，用于存储 shp 文件路径：

```csharp
//URL of the elevation service - provides elevation component of the scene.
string _elevation = "http://elevation3d.arcgis.com/arcgis/rest/services/WorldElevation3D/Terrain3D/ImageServer";
//URL of the building service - provides builidng models.
string _buildings = "https://tiles.arcgis.com/tiles/z2tnIkrLQ2BRzr6P/arcgis/rest/services/New_York_LoD2_3D_Buildings/SceneServer/layers/0";
//url of taxi model.
string _taxiModel = @"ArcGISRuntimeSampleData\dolmus_3ds\dolmus.3ds";
//Starting point of the observation point.
MapPoint _mpObserver = new MapPoint(-73.984988, 40.748131, 20, SpatialReferences.Wgs84);
//taxi point of the target point.
MapPoint _mpTaxi = new MapPoint(-73.984513, 40.748469, SpatialReferences.Wgs84);
//Graphic to represent the observation point.
Graphic _observer;
//Graphic to represent the observed target.
Graphic _taxi;
//Line of Sight Analysis.
GeoElementLineOfSight _gelos;
```

编写 Initialize 函数：

```csharp
void Initialize()
{
```

```csharp
//Create scene.
Scene scene = new Scene(Basemap.CreateImageryWithLabels())
{
    InitialViewpoint = new Viewpoint(_mpObserver, 1000000)
};
//Create the elevation source.
ElevationSource elevationSource = new ArcGISTiledElevationSource(new Uri(_elevation));
//Add the elevation source to the scene.
scene.BaseSurface.ElevationSources.Add(elevationSource);
//Create the building scene layer.
ArcGISSceneLayer sceneLayer = new ArcGISSceneLayer(new Uri(_buildings));
//Add the building layer to the scene.
scene.OperationalLayers.Add(sceneLayer);
//Add the scene to the view.
sceneView1.Scene = scene;
// Create a graphics overlay with relative surface placement; relative surface placement allows the Z position of the observation point to be adjusted.
GraphicsOverlay overlay = new GraphicsOverlay();
overlay.SceneProperties.SurfacePlacement = SurfacePlacement.Relative;
//Create the symbol that will symbolize the observation point.
SimpleMarkerSceneSymbol smss =new SimpleMarkerSceneSymbol(SimpleMarkerSceneSymbolStyle.Sphere, Color.Red, 10, 10, 10, SceneSymbolAnchorPosition.Bottom);
//Create the observation point graphic from the point and symbol.
_observer = new Graphic(_mpObserver, smss);
//Add the observer to the overlay.
overlay.Graphics.Add(_observer);
//Add the taxi to the scene.
string taxiPath = System.IO.Path.Combine(AppDomain.CurrentDomain.BaseDirectory + "..\\..\\..\\..\\", _taxiModel);
ModelSceneSymbol mss = await ModelSceneSymbol.CreateAsync(new Uri(taxiPath));
// Set the anchor position for the mode; ensures that the model appears above the ground.
mss.AnchorPosition = SceneSymbolAnchorPosition.Bottom;
```

```
    //Create the graphic from the taxi starting point and the symbol.
    _taxi = new Graphic(_mpTaxi, mss);
    //Add the taxi graphic to the overlay.
    overlay.Graphics.Add(_taxi);
    //Add the overlay to the scene.
    sceneView1.GraphicsOverlays.Add(overlay);
    //Create GeoElement Line of sight analysis (taxi to building).
    //Create the analysis.
    _gelos = new GeoElementLineOfSight(_observer, _taxi)
    {
        // Apply an offset to the target. This helps avoid some false negatives.
        TargetOffsetZ = 2
    };
    // Subscribe to TargetVisible events; allows for updating the UI and selecting the taxi when it is visible.
    _gelos.TargetVisibilityChanged += gelos_TargetVisibilityChanged;
    //Create the analysis overlay.
    AnalysisOverlay analysisOverlay = new AnalysisOverlay();
    //Add the analysis to the overlay.
    analysisOverlay.Analyses.Add(_gelos);
    //Add the analysis overlay to the scene.
    sceneView1.AnalysisOverlays.Add(analysisOverlay);
}
```

在构造函数中调用 Initialize 函数：
```
public MainWindow()
{
    InitializeComponent();
    Initialize();
}
```

点击场景，改变目标点，作为视线分析的终点：
```
void sceneView1_GeoViewTapped(object sender, GeoViewInputEventArgs e)
{
    _taxi.Geometry = e.Location;
}
```

更新目标可见性状态：
```
async void gelos_TargetVisibilityChanged(object sender, EventArgs e)
```

```
}
    await Dispatcher.BeginInvoke ( DispatcherPriority.Normal,
(Action)UpdateTargetVisibility);
}
async void gelos_TargetVisibilityChanged(object sender, EventArgs e)
{
    await Dispatcher.BeginInvoke ( DispatcherPriority.Normal,
(Action)UpdateTargetVisibility);
}
void UpdateTargetVisibility()
{
   tbLineOfSight.Text = $"GeoElementLineOfSight.TargetVisibility=
{_gelos.TargetVisibility.ToString()}";
   if (_gelos.TargetVisibility == LineOfSightTarget
Visibility.Visible)
       _taxi.IsSelected = true;
   else
       _taxi.IsSelected = false;
}
```

【实验结果】

待高程影像和三维建筑图层加载完成后，点击场景，设置观察目标，App 实时计算从观察点到目标点的通视性。可视部分用绿色表示，遮挡部分用红色表达。如果目标可见，则目标出租车处于选中状态，如果不可见(有遮挡)，目标出租车取消选中状态。

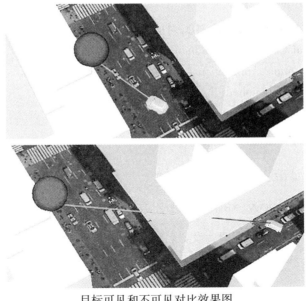

目标可见和不可见对比效果图

第 3 节　基于位置的三维视域分析

三维可视域分析可以理解为在监测点放置一个摄像机,计算其可以监控的范围。

摄像机可以根据多个参数进行调整,包括位置、方向角、俯仰角和可视的最小和最大距离。

观察位置:以 X、Y、Z 表示观察点的坐标值。

方向角:当前相机的方向与正北方向的夹角。

俯仰角:当前分析的相机方向与地平面的夹角。

可视距离:用来设置可视域分析时的长度范围,单位为米。

视域分析结果可以当作三维模型数据,通过布尔运算和空间位置关系进一步进行三维空间分析。

【实验目的】

在左边参数面板调整视域参数,在场景视图中点击视点,实时查看视域范围,其中绿色部分表示可见范围,红色部分表示遮挡范围。

【实验数据】

ArcGIS 在线高程服务:

http://elevation3d.arcgis.com/arcgis/rest/services/WorldElevation3D/Terrain3D/ImageServer

张家界天子山附近坐标:

MapPoint(110.4346619000, 29.3956642900, 1200)

【实验步骤】

1. 新建项目

.NET 框架:与 ArcGIS Runtime SDK 对应

模板列表:Visual C#/Windows/Windows Classic Desktop

模板:ArcGIS Runtime Application (WPF)

名称:LocationBasedViewshed

位置:D:\ArcGISRuntimeTutorial

2. 编译项目

详情见第 1 章第 3 节。

3. 修改 MainWindow.xaml 标记语言

移除地图视图,添加场景视图,设置名称属性:

`<esri:SceneView Name="mySceneView" />`

设计 UI:

第 3 节 基于位置的三维视域分析

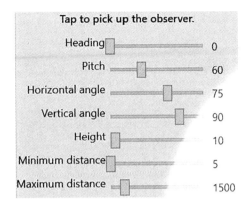

其对应 XAML 语言如下：

```
< Grid Background = " White " Opacity = " 0.8 " Width = " 250 " HorizontalAlignment = "Left" VerticalAlignment = "Top" >
    <Grid.RowDefinitions>
        <RowDefinition />
        <RowDefinition />
        <RowDefinition />
        <RowDefinition />
        <RowDefinition />
        <RowDefinition />
        <RowDefinition />
        <RowDefinition />
    </Grid.RowDefinitions>
    <Grid.ColumnDefinitions>
        <ColumnDefinition Width = "Auto" />
        <ColumnDefinition Width = " * " />
        <ColumnDefinition Width = "Auto" />
    </Grid.ColumnDefinitions>
    <TextBlock Text = "Tap to pick up the observer." Grid.Row = "0" Grid.Column = "0" Grid.ColumnSpan = "3" Margin = "0,0,0,10" TextWrapping = "WrapWithOverflow" TextAlignment = "Center" FontWeight = "SemiBold" />
     < TextBlock Text = " Heading " Grid.Row = " 1 " Grid.Column = " 0 " HorizontalAlignment = "Right" />
     < TextBlock Text = " Pitch " Grid.Row = " 2 " Grid.Column = " 0 " HorizontalAlignment = "Right" />
    <TextBlock Text = "Horizontal angle" Grid.Row = "3" Grid.Column = "0" HorizontalAlignment = "Right" />
     < TextBlock Text = "Vertical angle" Grid.Row = "4" Grid.Column = "0"
```

```xml
HorizontalAlignment="Right"/>
    <TextBlock Text="Height" Grid.Row="5" Grid.Column="0" HorizontalAlignment="Right"/>
    <TextBlock Text="Minimum distance" Grid.Row="6" Grid.Column="0" HorizontalAlignment="Right"/>
    <TextBlock Text="Maximum distance" Grid.Row="7" Grid.Column="0" HorizontalAlignment="Right"/>
    <Slider Name="HeadingSlider" Grid.Row="1" Grid.Column="1" HorizontalAlignment="Stretch" VerticalAlignment="Center" Value="0" Maximum="360" TickFrequency="1" IsSnapToTickEnabled="True" ValueChanged="HandleSettingsChange"/>
    <Slider Name="PitchSlider" Grid.Row="2" Grid.Column="1" HorizontalAlignment="Stretch" VerticalAlignment="Center" Value="60" Maximum="180" TickFrequency="1" IsSnapToTickEnabled="True" ValueChanged="HandleSettingsChange"/>
    <Slider Name="HorizontalAngleSlider" Grid.Row="3" Grid.Column="1" HorizontalAlignment="Stretch" VerticalAlignment="Center" Value="75" Maximum="120" Minimum="1" TickFrequency="1" IsSnapToTickEnabled="True" ValueChanged="HandleSettingsChange"/>
    <Slider Name="VerticalAngleSlider" Grid.Row="4" Grid.Column="1" HorizontalAlignment="Stretch" VerticalAlignment="Center" Value="90" Maximum="120" Minimum="1" TickFrequency="1" IsSnapToTickEnabled="True" ValueChanged="HandleSettingsChange"/>
    <Slider Name="HeightSlider" Grid.Row="5" Grid.Column="1" HorizontalAlignment="Stretch" VerticalAlignment="Center" Value="10" Maximum="200" Minimum="0" TickFrequency="1" IsSnapToTickEnabled="True" ValueChanged="HandleSettingsChange"/>
    <Slider Name="MinimumDistanceSlider" Grid.Row="6" Grid.Column="1" HorizontalAlignment="Stretch" VerticalAlignment="Center" Value="5" Maximum="8999" Minimum="5" TickFrequency="1" IsSnapToTickEnabled="True" ValueChanged="HandleSettingsChange"/>
    <Slider Name="MaximumDistanceSlider" Grid.Row="7" Grid.Column="1" HorizontalAlignment="Stretch" VerticalAlignment="Center" Value="1500" Minimum="1" Maximum="9999" TickFrequency="1" IsSnapToTickEnabled="True" ValueChanged="HandleSettingsChange"/>
    <Label Content="{Binding Value, ElementName=HeadingSlider}" Grid.Row="1" Grid.Column="2"/>
    <Label Content="{Binding Value, ElementName=PitchSlider}"
```

```
Grid.Row="2" Grid.Column="2"/>
    <Label Content="{Binding Value, ElementName=Horizontal
AngleSlider}" Grid.Row="3" Grid.Column="2"/>
    <Label Content="{Binding Value, ElementName=VerticalAngle
Slider}" Grid.Row="4" Grid.Column="2"/>
    <Label Content="{Binding Value, ElementName=HeightSlider}"
Grid.Row="5" Grid.Column="2"/>
    <Label Content="{Binding Value, ElementName=MinimumDistance
Slider}" Grid.Row="6" Grid.Column="2"/>
    <Label Content="{Binding Value, ElementName=MaximumDistance
Slider}" Grid.Row="7" Grid.Column="2"/>
</Grid>
```

4. 添加程序集引用：

System.Drawing

5. 修改 MainWindow.xaml.cs 代码

添加命名空间：

```
using Esri.ArcGISRuntime.Geometry;
using Esri.ArcGISRuntime.Mapping;
using Esri.ArcGISRuntime.Symbology;
using Esri.ArcGISRuntime.UI;
using Esri.ArcGISRuntime.UI.Controls;
using Esri.ArcGISRuntime.UI.GeoAnalysis;
using System;
using System.Drawing;
using System.Windows;
using System.Windows.Input;
```

生成全局变量：

```
//Hold the URL to the elevation source.
private readonly Uri _elevationUri = new Uri("http://elevation3d.
arcgis.com/arcgis/rest/services/WorldElevation3D/Terrain3D/
ImageServer");
//Hold a reference to the viewshed analysis.
private LocationViewshed _viewshed;
//Hold a reference to the analysis overlay that will hold the viewshed
analysis.
private AnalysisOverlay _analysisOverlay;
```

```
Graphic _viewPointGraphic;
//Graphics overlay for viewpoint symbol.
private GraphicsOverlay _viewpointOverlay;
//Height of the viewpoint above the ground.
private double _viewHeight;
```
编写 Initialize 函数：
```
void Initialize()
{
    _viewHeight = HeightSlider.Value;
    //Create the scene with the imagery basemap.
    Scene myScene = new Scene(Basemap.CreateImageryWithLabelsVector());
    mySceneView.Scene = myScene;
    //Add the surface elevation.
    Surface mySurface = new Surface();
    mySurface.ElevationSources.Add(new ArcGISTiledElevationSource(_elevationUri));
    myScene.BaseSurface = mySurface;
    //Create the MapPoint representing the initial location.
    MapPoint initialLocation = new MapPoint(110.4346619000, 29.3956642900, 1200 + _viewHeight);
    //Create the location viewshed analysis.
    _viewshed = new LocationViewshed(
        initialLocation,
        HeadingSlider.Value,
        PitchSlider.Value,
        HorizontalAngleSlider.Value,
        VerticalAngleSlider.Value,
        MinimumDistanceSlider.Value,
        MaximumDistanceSlider.Value);
    _viewshed.IsVisible = true;
    _viewshed.IsFrustumOutlineVisible = true;
    //Create a camera based on the initial location.
    Camera camera = new Camera(initialLocation, 2000.0, 20.0, 70.0, 0.0);
    //Create a symbol for the viewpoint.
    SimpleMarkerSceneSymbol viewpointSymbol = SimpleMarkerSceneSymbol.CreateSphere(Color.Blue, 10, SceneSymbolAnchorPosition.
```

```
Center);
    //Add the symbol to the viewpoint overlay.
    _viewpointOverlay = new GraphicsOverlay
    {
        SceneProperties = new LayerSceneProperties(SurfacePlacement.Absolute)
    };
    _viewPointGraphic = new Graphic(initialLocation, viewpointSymbol);
    _viewpointOverlay.Graphics.Add(_viewPointGraphic);
    //Apply the camera to the scene view.
    mySceneView.SetViewpointCamera(camera);
    //Create an analysis overlay for showing the viewshed analysis.
    _analysisOverlay = new AnalysisOverlay();
    //Add the viewshed analysis to the overlay.
    _analysisOverlay.Analyses.Add(_viewshed);
    //Add the analysis overlay to the SceneView.
    mySceneView.AnalysisOverlays.Add(_analysisOverlay);
    //Add the graphics overlay.
    mySceneView.GraphicsOverlays.Add(_viewpointOverlay);
    // Subscribe to tap events. This enables the 'pick up' and 'drop' workflow for moving the viewpoint.
    mySceneView.GeoViewTapped += mySceneViewOnGeoViewTapped;
}
```

在构造函数中调用 Initialize 函数：

```
public MainWindow()
{
    InitializeComponent();
    Initialize ();
}
```

点击场景，设置当前视点：

```
void mySceneViewOnGeoViewTapped(object sender, GeoViewInputEventArgs e)
{
    //Get the corresponding MapPoint.
    MapPoint mapLocation = e.Location;
    //Return if the MapPoint is null. This might happen if mouse leaves SceneView area.
```

```
    if (e.Location == null)
    {
        return;
    }
    // Adjust the Z value of the MapPoint to reflect the selected
height.
    mapLocation = new MapPoint (mapLocation.X, mapLocation.Y,
mapLocation.Z + _viewHeight);
    //Update the viewshed.
    _viewshed.Location = mapLocation;
    //Update the viewpoint symbol.
    _viewPointGraphic.Geometry = mapLocation;
}
```

用户调整视域分析参数：

```
void mySceneViewOnGeoViewTapped(object sender, GeoViewInput
EventArgs e)
{
    //Get the corresponding MapPoint.
    MapPoint mapLocation = e.Location;
    //Return if the MapPoint is null. This might happen if mouse leaves
SceneView area.
    if (e.Location == null)
        return;
    // Adjust the Z value of the MapPoint to reflect the selected
height.
    mapLocation = new MapPoint (mapLocation.X, mapLocation.Y,
mapLocation.Z + _viewHeight);
    //Update the viewshed.
    _viewshed.Location = mapLocation;
    //Update the viewpoint symbol.
    _viewPointGraphic.Geometry = mapLocation;
}
```

【实验结果】

在左边参数面板调整视域参数，在场景视图中点击视点，实时查看视域范围，其中绿色部分表示可见范围，红色部分表示遮挡范围。

第4节 基于服务的最近设施分配

最近设施分配是一种路径分析，在网络上指定一个事件点和一组设施点，为事件点查找以最小耗费能到达的一个或几个设施点，结果为从事件点到设施点（或从设施点到事件点）的最佳路径。

最近设施分配应用在应急救援、物流配送等领域。例如发生交通事故后，要求查找在8分钟内能到达的最近医院。事故发生地是事件点，周边医院则是设施点。应用路径分析时也可以设置障碍边和障碍点，在行驶路径上这些障碍将不能被穿越。

【实验目的】

为每个事件点 incidents，分配最近的设施 facility。

【实验数据】

ArcGIS Online 在线底图服务：

https：//sampleserver6.arcgisonline.com/arcgis/rest/services/World_Street_Map/MapServer

本地 shp 文件，内容为小比例尺的中国各省级行政区划边界：ChinaProvince \ bou2_4p.shp

【实验步骤】

1. 新建项目

.NET 框架：与 ArcGIS Runtime SDK 对应
模板列表：Visual C#/Windows/Windows Classic Desktop
模板：ArcGIS Runtime Application（WPF）
名称：ServiceStaticClosestFacility
位置：D：\ArcGISRuntimeTutorial

2. 编译应用程序

详情见第1章第3节。

3. 修改 MainWindow.xaml 标记语言

对地图视图，添加名称属性：

```
<esri:MapView Name="mapView1" Map="{Binding Map, Source={Static ResourceMapViewModel}}"/>
```

设计 UI：

```
<Button Name="SolveRoutesButton" Content="Solve Routes" Click="SolveRoutesClick"/>
```

4. 修改 MainWindow.xaml.cs 代码

添加命名空间：

```
using Esri.ArcGISRuntime.Data;
using Esri.ArcGISRuntime.Geometry;
using Esri.ArcGISRuntime.Mapping;
using Esri.ArcGISRuntime.Symbology;
using Esri.ArcGISRuntime.Tasks.NetworkAnalysis;
using Esri.ArcGISRuntime.UI;
```

生成全局变量，用于存储服务路径：

```
//Table of all facilities.
ServiceFeatureTable _facilityTable;
//Table of all incidents.
ServiceFeatureTable _incidentTable;
//Uri for facilities feature service.
Uri _facilityUri = new Uri("https://services2.arcgis.com/ZQgQTuoyBrtmoGdP/ArcGIS/rest/services/San_Diego_Facilities/FeatureServer/0");
//Uri for incident feature service.
Uri _incidentUri = new Uri("https://services2.arcgis.com/ZQgQTuoyBrtmoGdP/ArcGIS/rest/services/San_Diego_Incidents/FeatureServer/0");
//Uri for the closest facility service.
Uri _closestFacilityUri = new Uri("http://sampleserver6.arcgisonline.com/arcgis/rest/services/NetworkAnalysis/SanDiego/NAServer/ClosestFacility");
```

编写 Initialize 函数：

```
async void Initialize()
{
    //Create a symbol for displaying facilities.
```

```csharp
    PictureMarkerSymbol facilitySymbol = new PictureMarkerSymbol
(new Uri("http://static.arcgis.com/images/Symbols/SafetyHealth/
FireStation.png"));
    //Incident symbol.
    PictureMarkerSymbol incidentSymbol = new PictureMarkerSymbol
(new Uri("http://static.arcgis.com/images/Symbols/SafetyHealth/
esriCrimeMarker_56_Gradient.png"));
    //Create a table for facilities using the FeatureServer.
    _facilityTable = new ServiceFeatureTable(_facilityUri);
    //Create a feature layer from the table.
    FeatureLayer facilityLayer = new FeatureLayer(_facilityTable)
    {
        Renderer = new SimpleRenderer(facilitySymbol)
    };
    //Create a table for facilities using the FeatureServer.
    _incidentTable = new ServiceFeatureTable(_incidentUri);
    //Create a feature layer from the table.
    FeatureLayer incidentLayer = new FeatureLayer(_incidentTable)
    {
        Renderer = new SimpleRenderer(incidentSymbol)
    };
    //Add the layers to the map.
    mapView1.Map.OperationalLayers.Add(facilityLayer);
    mapView1.Map.OperationalLayers.Add(incidentLayer);
    //Wait for both layers to load.
    await facilityLayer.LoadAsync();
    await incidentLayer.LoadAsync();
    //Zoom to the combined extent of both layers.
    Envelope fullExtent = GeometryEngine.CombineExtents(facility
Layer.FullExtent, incidentLayer.FullExtent);
    await mapView1.SetViewpointGeometryAsync(fullExtent, 50);
}
```

在构造函数中调用 Initialize 函数：

```csharp
public MainWindow()
{
    InitializeComponent();
    Initialize();
}
```

最近设施分配：

```
async void SolveRoutesClick(object sender, EventArgs e)
{
    QueryParameters queryParams = new QueryParameters()
    {WhereClause="1=1"};
    //Query all features in the facility table.
    FeatureQueryResult facilityResult = await _facilityTable.QueryFeaturesAsync(queryParams);
    // Add all of the query results to facilities as new Facility objects.
    var facilities = facilityResult.ToList().Select(feature => new Facility((MapPoint)feature.Geometry));
    //Query all features in the incident table.
    FeatureQueryResult incidentResult = await _incidentTable.QueryFeaturesAsync(queryParams);
    // Add all of the query results to facilities as new Incident objects.
    var incidents = incidentResult.ToList().Select(feature => new Incident((MapPoint)feature.Geometry)).ToList();
    // Solves task to find closest route between an incident and a facility.
    //Create a ClosestFacilityTask using the San Diego Uri.
    ClosestFacilityTask task = await ClosestFacilityTask.CreateAsync(_closestFacilityUri);
    //Set facilities and incident in parameters.
    ClosestFacilityParameters closestFacilityParameters = await task.CreateDefaultParametersAsync();
    closestFacilityParameters.SetFacilities(facilities);
    closestFacilityParameters.SetIncidents(incidents);
    //Use the task to solve for the closest facility.
    ClosestFacilityResult result = await task.SolveClosestFacilityAsync(closestFacilityParameters);
    // Add a graphics overlay to MyMapView. (Will be used later to display routes)
    mapView1.GraphicsOverlays.Add(new GraphicsOverlay());
    for (int i = 0; i < incidents.Count; i++)
    {
        //Get the index of the closest facility to incident. (i) is the
```

index of the incident, [0] is the index of the closest facility.
        int closestFacility = result.GetRankedFacilityIndexes(i)[0];
        //Get the route to the closest facility.
        ClosestFacilityRoute route = result.GetRoute(closestFacility, i);
        //Display the route on the graphics overlay.
            mapView1.GraphicsOverlays[0].Graphics.Add(new Graphic(route.RouteGeometry));
    }
}
```

**【实验结果】**

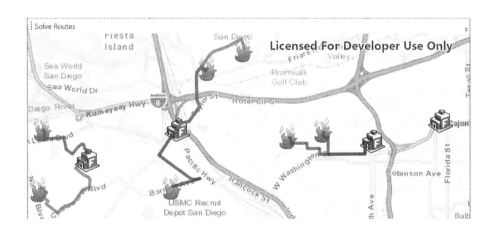

## 第 5 节 基于本地数据的最近设施分配

利用本地数据进行分析，可以在离线状态下进行最近设施分配，从而增加灵活性。

**【实验目的】**

为每个事件点 incidents，分配最近的设施 facility。数据来自本地 geodatabase。

**【实验数据】**

本地服务设施图标(消防站)：
"ArcGISRuntimeSampleData\images\Symbols\SafetyHealth\FireStation.png";

本地实践图标(火灾)：
"ArcGISRuntimeSampleData\images\Symbols\SafetyHealth\esriCrimeMarker_56_Gradient.png"

本地移动地理数据库(geodatabase)：
"ArcGISRuntimeSampleData\network\notDissolved\geodatabase.
```

geodatabase"

【实验步骤】

1. 新建项目

.NET 框架：与 ArcGIS Runtime SDK 对应
模板列表：Visual C#/Windows/Windows Classic Desktop
模板：ArcGIS Runtime Application（WPF）
名称：LocalStaticClosestFacility
位置：D：\ArcGISRuntimeTutorial

2. 编译项目

详情见第 1 章第 3 节。

3. 修改 MainWindow.xaml 标记语言

对地图视图，添加名称属性：

```
<esri:MapView Name="mapView1" Map="{Binding Map, Source={StaticResource MapViewModel}}" />
```

设计 UI：

```
<StackPanel VerticalAlignment="Top">
    <TextBox Name="tbGeodatabase" />
    <Button Name="SolveRoutesButton" Content="Solve Routes" Click="SolveRoutesClick" />
</StackPanel>
```

4. 修改 MainWindow.xaml.cs 代码

添加命名空间：

```
using Esri.ArcGISRuntime.Data;
using Esri.ArcGISRuntime.Geometry;
using Esri.ArcGISRuntime.Mapping;
using Esri.ArcGISRuntime.Symbology;
using Esri.ArcGISRuntime.Tasks.NetworkAnalysis;
using Esri.ArcGISRuntime.UI;
using System.Drawing;
```

生成全局变量，用于存储数据路径。
本地服务设施图标（消防站）：

```
string _FireStation = @"ArcGISRuntimeSampleData\images\Symbols\SafetyHealth\FireStation.png";
```

本地事件图标（火灾）：

```csharp
string _esriCrimeMarker_56_Gradient = @"ArcGISRuntimeSampleData\
images\Symbols\SafetyHealth\esriCrimeMarker_56_Gradient.png";
```
本地移动地理数据库(geodatabase)：
```csharp
string _geodatabasePath = @"ArcGISRuntimeSampleData\network\
notDissolved\geodatabase.geodatabase";
```
编写 Initialize 函数：
```csharp
void Initialize()
{
    //Create a symbol for displaying facilities.
    PictureMarkerSymbol facilitySymbol = new PictureMarkerSymbol
(new Uri(System.IO.Path.Combine(AppDomain.CurrentDomain.Base
Directory + "..\\..\\..\\..\\", _FireStation)));
    //Incident symbol.
    PictureMarkerSymbol incidentSymbol = new PictureMarkerSymbol
(new Uri(System.IO.Path.Combine(AppDomain.CurrentDomain.Base
Directory + "..\\..\\..\\..\\", _esriCrimeMarker_56_Gradient)));
    tbGeodatabase.Text = System.IO.Path.Combine(AppDomain.Current
Domain.BaseDirectory + "..\\..\\..\\..\\", _geodatabasePath);
    //Get the new geodatabase
    Geodatabase gdb = await Geodatabase.OpenAsync(tbGeodatabase.
Text);
    if (mapView1.Map == null)
        mapView1.Map = new Map();
    //Loop through all feature tables in the geodatabase and add a new
layer to the map.
    foreach (GeodatabaseFeatureTable table in gdb.GeodatabaseFeature
Tables)
    {
        FeatureLayer layer = new FeatureLayer(table);
        await layer.LoadAsync();
        if (table.TableName == "park")
        {
            _facilityTable = table;
            layer.Renderer = new SimpleRenderer(facilitySymbol);
        }
        else if (table.TableName == "point")
        {
            _incidentTable = table;
```

```
            layer.Renderer = new SimpleRenderer(incidentSymbol);
        }
        mapView1.Map.OperationalLayers.Add(layer);
    }
    mapView1.ZoomToOperationalLayrAsync();
}
```

在构造函数中调用 Initialize 函数：

```
public MainWindow()
{
    InitializeComponent();
    Initialize();
}
```

最近设施分配：

```
async void SolveRoutesClick(object sender, EventArgs e)
{
    QueryParameters queryParams = new QueryParameters()
    {
        WhereClause = "1=1"
    };
    //Query all features in the facility table.
    FeatureQueryResult facilityResult = await _facilityTable.QueryFeaturesAsync(queryParams);
    // Add all of the query results to facilities as new Facility objects.
    var facilities = facilityResult.ToList().Select(feature => new Facility((MapPoint)feature.Geometry));
    //Query all features in the incident table.
    FeatureQueryResult incidentResult = await _incidentTable.QueryFeaturesAsync(queryParams);
    // Add all of the query results to facilities as new Incident objects.
    var incidents = incidentResult.ToList().Select(feature => new Incident((MapPoint)feature.Geometry)).ToList();
    ClosestFacilityTask task = await ClosestFacilityTask.CreateAsync(tbGeodatabase.Text, "ds_ND");
    //Set facilities and incident in parameters.
    ClosestFacilityParameters closestFacilityParameters = await task.CreateDefaultParametersAsync();
```

```
closestFacilityParameters.SetFacilities(facilities);
closestFacilityParameters.SetIncidents(incidents);
//Use the task to solve for the closest facility.
ClosestFacilityResult result = await task.SolveClosestFacilityAsync(closestFacilityParameters);
// Add a graphics overlay to MyMapView. (Will be used later to display routes)
mapView1.GraphicsOverlays.Add(new GraphicsOverlay());
for (int i = 0; i < incidents.Count; i++)
{
    //Get the index of the closest facility to incident. (i) is the index of the incident, [0] is the index of the closest facility.
    int closestFacility = result.GetRankedFacilityIndexes(i)[0];
    //Get the route to the closest facility.
    ClosestFacilityRoute route = result.GetRoute(closestFacility, i);
    //Display the route on the graphics overlay.
    mapView1.GraphicsOverlays[0].Graphics.Add(new Graphic(route.RouteGeometry, new SimpleLineSymbol(SimpleLineSymbolStyle.Solid, Color.Red, 2)));
}
}
```

【实验结果】

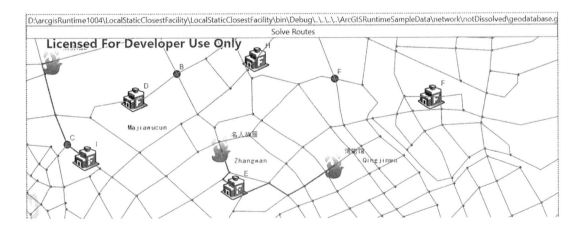

第 6 节　本地数据的服务范围分配

服务区分析设施的服务范围，可采用在城市规划和区域规划中公共设施的服务范围和辐射范围。除考虑距离和空间可达性之外，还考虑公共服务设施本身的吸引力和服务能力。服务范围分配时，常采用引力模型来计算距离和城市本身的指标。

【实验目的】

在本地地理数据库中存有交通网络数据集，用户交互规划服务设施，放置障碍模拟道路临时封闭，动态计算不同规划条件下的设施服务范围。

【实验数据】

本地交通网络数据集：ArcGISRuntimeSampleData \ network \ notDissolved \ geodatabase.geodatabase

【实验步骤】

1. 新建项目

.NET 框架：与 ArcGIS Runtime SDK 对应

模板列表：Visual C#/Windows/Windows Classic Desktop

模板：ArcGIS Runtime Application（WPF）

名称：ServiceArea

位置：D：\ArcGISRuntimeTutorial

2. 编译项目

详情见第 1 章第 3 节。

3. 修改 MainWindow.xaml 标记语言

对地图视图，添加名称属性：

```
<esri:MapView Name="mapView1" Map="{Binding Map,Source={StaticResource MapViewModel}}" />
```

设计 UI：

```
ampleData\network\notDissolved\geodatabase.geodatabase
Place facility  Draw barrier  Show service areas  Reset
```

```
<StackPanel VerticalAlignment="Top">
    <TextBox Name="tbGeodatabase" VerticalAlignment="Top" />
    <ToolBar>
```

```xml
        <Button Name="PlaceFacilityButton" Content="Place facility" Click="PlaceFacilityButton_Click"/>
         <Button Name="DrawBarrierButton" Content="Draw barrier" Click="DrawBarrierButton_Click"/>
          <Button Name="ShowServiceAreasButton" Content="Show service areas" Click="ShowServiceAreasButton_Click"/>
           <Button Name="ResetButton" Content="Reset" Click="Reset_Click"/>
    </ToolBar>
</StackPanel>
```

4. 修改 MainWindow.xaml.cs 代码

添加命名空间：

```
using Esri.ArcGISRuntime.Data;
using Esri.ArcGISRuntime.Geometry;
using Esri.ArcGISRuntime.Mapping;
using Esri.ArcGISRuntime.Symbology;
using Esri.ArcGISRuntime.Tasks.NetworkAnalysis;
using Esri.ArcGISRuntime.UI;
using System.Drawing;
```

生成全局变量，存储数据路径：

```
string _geodatabasePath = @"ArcGISRuntimeSampleData\network\notDissolved\geodatabase.geodatabase";
string _Hospital = @"ArcGISRuntimeSampleData\images\Symbols\SafetyHealth\Hospital.png";
```

编写 Initialize 函数：

```
async void Initialize()
{
    tbGeodatabase.Text = System.IO.Path.Combine(AppDomain.CurrentDomain.BaseDirectory + "..\\..\\..\\..\\", _geodatabasePath);
    //Create a symbol for displaying facilities.
     PictureMarkerSymbol facilitySymbol = new PictureMarkerSymbol(new Uri(System.IO.Path.Combine(AppDomain.CurrentDomain.BaseDirectory + "..\\..\\..\\..\\", _Hospital)));
    //route symbol
    SimpleLineSymbol lineSymbol = new SimpleLineSymbol(Simple
```

```
LineSymbolStyle.Solid, Color.FromArgb(255, 0, 0, 255), 5.0f);
    //Get the new geodatabase.
    Geodatabase gdb = await Geodatabase.OpenAsync(tbGeodatabase.Text);
    if (mapView1.Map == null)
        mapView1.Map = new Map();
    //Loop through all feature tables in the geodatabase and add a new layer to the map.
    foreach (GeodatabaseFeatureTable table in gdb.GeodatabaseFeatureTables)
    {
        FeatureLayer layer = new FeatureLayer(table);
        await layer.LoadAsync();
        mapView1.Map.OperationalLayers.Add(layer);
    }
    mapView1.ZoomToOperationalLayrAsync();
    //GraphicsOverlays[0]: facility
    mapView1.GraphicsOverlays.Add(new GraphicsOverlay()
        {Renderer = new SimpleRenderer(facilitySymbol)}
        );
    //GraphicsOverlays[1]: barrier
    mapView1.GraphicsOverlays.Add ( new  GraphicsOverlay ( ) {
Renderer = new SimpleRenderer(lineSymbol) });
    //GraphicsOverlays[2]: service area
    mapView1.GraphicsOverlays.Add(new GraphicsOverlay());
    //Add a new behavior for double taps on the MapView.
    mapView1.GeoViewDoubleTapped += (s, e) =>
    {
        //If the sketch editor complete command is enabled, a sketch is in progress.
        if (mapView1.SketchEditor.CompleteCommand.CanExecute(null))
            {e.Handled = true;}
    };
    //Create the service area task and parameters based on the Uri.
    _task = await ServiceAreaTask.CreateAsync(tbGeodatabase.Text, "ds_ND");
```

在构造函数中调用 Initialize 函数:
```
public MainWindow()
{
    InitializeComponent();
    Initialize();
}
```
放置服务设施:
```
private async void PlaceFacilityButton_Click(object sender, RoutedEventArgs e)
{
    //Let the user tap on the map view using the point sketch mode.
    SketchCreationMode creationMode = SketchCreationMode.Point;
    Geometry geometry = await mapView1.SketchEditor.StartAsync(creationMode, false);
    //Create a graphic for the facility.
    Graphic facilityGraphic = new Graphic(geometry);
    //Add the graphic to the graphics overlay.
    mapView1.GraphicsOverlays[0].Graphics.Add(facilityGraphic);
}
```
绘制障碍:
```
async void DrawBarrierButton_Click(object sender, RoutedEventArgs e)
{
    //Let the user draw on the map view using the polyline sketch mode.
    SketchCreationMode creationMode = SketchCreationMode.Polyline;
    Geometry geometry = await mapView1.SketchEditor.StartAsync(creationMode, false);
    //Symbol for the barriers.
    SimpleLineSymbol barrierSymbol = new SimpleLineSymbol(SimpleLineSymbolStyle.Solid, Color.Black, 5.0f);
    //Create the graphic to be used for barriers.
    Graphic barrierGraphic = new Graphic(geometry, barrierSymbol);
    //Add a graphic from the polyline the user drew.
    mapView1.GraphicsOverlays[1].Graphics.Add(barrierGraphic);
}
```
计算服务范围并显示:

第 8 章 地理信息处理服务

```
async void ShowServiceAreasButton_Click(object sender, RoutedEvent
Args e)
{
    var serviceAreaFacilities = from g in mapView1.GraphicsOverlays
[0].Graphics select new ServiceAreaFacility((MapPoint)g.Geometry);
    //Check that there is at least 1 facility to find a service area
for.
    if (!serviceAreaFacilities.Any())
    {
        MessageBox.Show("Must have at least one Facility!", "error");
    return;
    }
    //Store the default parameters for the service area in an object.
    ServiceAreaParameters serviceAreaParameters = await _task.Create
DefaultParametersAsync();
    //Add impedance cutoffs for facilities (drive time minutes).
    serviceAreaParameters.DefaultImpedanceCutoffs.Add(200);
  serviceAreaParameters.DefaultImpedanceCutoffs.Add(500);
   //Set the level of detail for the polygons.
    serviceAreaParameters.PolygonDetail = ServiceAreaPolygonDetail.
High;
    //Get a list of the barriers from the graphics overlay.
    var polylineBarriers = from g in mapView1.GraphicsOverlays[1]
.Graphics select new PolylineBarrier((Polyline)g.Geometry);
    //Add the barriers to the service area parameters.
    serviceAreaParameters.SetPolylineBarriers(polylineBarriers);
    //Update the parameters to include all of the placed facilities.
    serviceAreaParameters.SetFacilities(serviceAreaFacilities);
    //Clear existing graphics for service areas.
    mapView1.GraphicsOverlays[2].Graphics.Clear();
    //Solve for the service area of the facilities.
    ServiceAreaResult result = await _task.SolveServiceAreaAsync
(serviceAreaParameters);
    //Loop over each facility.
    for (int i = 0; i < serviceAreaFacilities.Count(); i++)
    {
```

```csharp
        //Create list of polygons from a service facility.
        List<ServiceAreaPolygon> polygons = result.GetResultPolygons(i).ToList();
        //Symbol for the outline of the service areas.
        SimpleLineSymbol serviceOutline = new SimpleLineSymbol(SimpleLineSymbolStyle.Solid, Color.LightGreen, 3.0f);
        //Create a list of fill symbols for the polygons.
        List<SimpleFillSymbol> fillSymbols = new List<SimpleFillSymbol>();
        fillSymbols.Add(new SimpleFillSymbol(SimpleFillSymbolStyle.Solid, Color.FromArgb(70, 255, 0, 0), serviceOutline));
        fillSymbols.Add(new SimpleFillSymbol(SimpleFillSymbolStyle.Solid, Color.FromArgb(70, 255, 165, 0), serviceOutline));
        //Loop over every polygon in every facilities result.
        for (int j = 0; j < polygons.Count; j++)
        {
            // Create the graphic for the service areas, alternating between fill symbols.
            Graphic serviceGraphic = new Graphic(polygons[j].Geometry, fillSymbols[j % 2]);
            // Graphic serviceGraphic = new Graphic(polygons[j].Geometry, fillSymbols[0]);
            //Add graphic for service area. Alternate the color of each polygon.
            mapView1.GraphicsOverlays[2].Graphics.Add(serviceGraphic);
        }
    }
}
```

点击重置:
```csharp
void Reset_Click(object sender, RoutedEventArgs e)
{
    //Clear all of the graphics.
    mapView1.GraphicsOverlays[0].Graphics.Clear();
    mapView1.GraphicsOverlays[1].Graphics.Clear();
    mapView1.GraphicsOverlays[2].Graphics.Clear();
}
```

【实验结果】

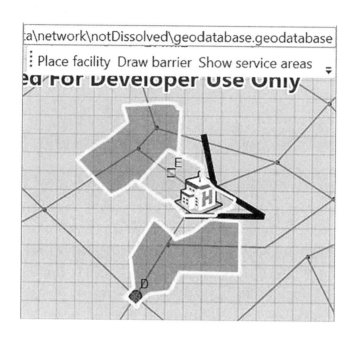

第7节　等高线生成器

等高线是地形图上高程相等的相邻各点所连成的闭合曲线。传统的地图使用者比较习惯使用等高线。等高线可以认为是数字高程模型的衍生产品。

创建等高线，需要使用本地地理处理包(geoprocessing package)，作为地图服务器。地图服务器是 ArcGIS Runtime 的扩展，可以提供更为丰富的、高级的地理处理和分析功能。

【实验目的】

通过本地地理服务，创建等高线。

【实验数据】

ArcGISRuntimeSampleData \ Contour_gpk \ Contour.gpk

【实验步骤】

1. 新建项目

.NET 框架：与 ArcGIS Runtime SDK 对应

模板列表：Visual C#/Windows/Windows Classic Desktop

模板：ArcGIS Runtime Application (WPF)

名称：ContourGenerator

位置：D:\ArcGISRuntimeTutorial

2. 编译项目

详情见第 1 章第 3 节。

3. 添加 LocalServer SDk

点击菜单 Project/Manage NuGet packages：

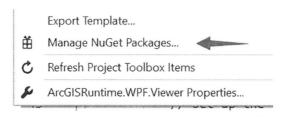

在 Browse 选项卡中，输入 Esri.ArcGISRuntime.LocalServices，找到对应版本的安装包后，点击右边的 Install 进行安装：

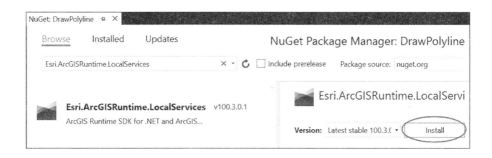

4. 修改 MainWindow.xaml 标记语言

对地图视图，添加名称属性：
<esri:MapView Name=" mapView1" Map=" {Binding Map, Source ={StaticResource MapViewModel}}" />

设计 UI：
```
<StackPanel Opacity="0.8">
    <TextBox Name="tbTpk" />
    <TextBox Name="tbGpk" />
    <ProgressBar Name="progressBar1" Height="20"></ProgressBar>
    <ToolBar>
        <Label Content="ContourInterval" />
```

```xml
<TextBox Name = "tbContourInterval" />
<Label Content = "m" />
<Separator />
<Button Name = "btGenerateContour" Content = "GenerateContour" Click = "btGenerateContour_Click" />
    <Button Name = "btReset" Content = "Reset" Click = "btReset_Click" />
</ToolBar>
 <TextBox Name = "tbStatus" Width = "300" HorizontalAlignment = "Left" />
</StackPanel>
```

5. 修改 MainWindow.xaml.cs 代码

添加命名空间：

```csharp
using Esri.ArcGISRuntime.LocalServices;
using Esri.ArcGISRuntime.Mapping;
using Esri.ArcGISRuntime.Tasks;
using Esri.ArcGISRuntime.Tasks.Geoprocessing;
using Esri.ArcGISRuntime.Data;
using System.IO;
```

生成全局变量，用于存储本地切片包和地理处理包路径：

```csharp
string _file = @"ArcGISRuntimeSampleData\RasterHillshade_tpk\RasterHillshade.tpk";
string _gp = @"ArcGISRuntimeSampleData\Contour_gpk\Contour.gpk";
```

定义本地服务相关变量：

```csharp
//Hold a reference to the local geoprocessing service.
LocalGeoprocessingService _lgpService;
//Hold a reference to the task:
GeoprocessingTask _gpTask;
//Hold a reference to the job:
GeoprocessingJob _gpJob;
```

编写 InitUI 函数：

```csharp
void InitUI()
{
    string rasterPath = System.IO.Path.Combine(AppDomain.CurrentDomain.BaseDirectory + "..\\..\\..\\..\\", _file);
```

```csharp
        tbTpk.Text = rasterPath;
        string gpServiceUrl = System.IO.Path.Combine(AppDomain.CurrentDomain.BaseDirectory + "..\\..\\..\\..\\", _gp);
        tbGpk.Text = gpServiceUrl;
        tbContourInterval.Text = "200";
}
```

添加底图：
```csharp
async void AddBaseMap()
{
    //Create a tile cache using the path to the raster.
    TileCache tileCache = new TileCache(tbTpk.Text);
    //Create the tiled layer from the tile cache.
    ArcGISTiledLayer tiledLayer = new ArcGISTiledLayer(tileCache);
    //Wait for the layer to load.
    await tiledLayer.LoadAsync();
    if (mapView1.Map == null)
        mapView1.Map = new Map();
    else
    //Zoom to extent of the tiled layer.
        await mapView1.SetViewpointGeometryAsync(tiledLayer.FullExtent);
    //Add the layer to the map.
        mapView1.Map.OperationalLayers.Add(tiledLayer);
}
```

在构造函数中调用 Initialize 函数：
```csharp
public MainWindow()
{
    InitializeComponent();
    InitUI();
    AddBaseMap();
}
```

生成等高线按钮响应事件：
```csharp
async void btGenerateContour_Click(object sender, RoutedEventArgs e)
{
    LocalServer.Instance.StatusChanged +=
    (s, args) => tbStatus.Text += $"LocalServer.Instance.Status =
```

```
{args.Status}\n";
    //Start the local server instance.
    await LocalServer.Instance.StartAsync();
    //Create the geoprocessing service.
    _lgpService = new LocalGeoprocessingService(tbGpk.Text,
GeoprocessingServiceType.AsynchronousSubmitWithMapServiceResult);
    //Take action once the service loads.
    _lgpService.StatusChanged += (s, args) => tbStatus.Text += $"Local
GeoprocessingService.Status={args.Status}\n";
    //Try to start the service.
    await _lgpService.StartAsync();
    //Create the geoprocessing task from the service.
    _gpTask = await GeoprocessingTask.CreateAsync(new Uri(_lgp
Service.Url + "/Contour"));
    //Create the geoprocessing parameters.
    GeoprocessingParameters gpParams = new GeoprocessingParameters
(GeoprocessingExecutionType.AsynchronousSubmit);
    //Add the interval parameter to the geoprocessing parameters.
    double.TryParse(tbContourInterval.Text, out double contour
Interval);
    gpParams.Inputs["ContourInterval"] = new GeoprocessingDouble
(contourInterval);
    //Create the job.
    _gpJob = _gpTask.CreateJob(gpParams);
    //Update the UI when job progress changes.
    _gpJob.ProgressChanged += (s, args) =>
    {
        Dispatcher.Invoke(() =>
        {
            progressBar1.Value = _gpJob.Progress;
        });
    };
    //Be notified when the task completes (or other change happens).
    _gpJob.JobChanged += GpJobOnJobChanged;
    //Start the job.
    _gpJob.Start();
}
```

添加地理处理结果：
```csharp
async void GpJobOnJobChanged(object o, EventArgs eventArgs)
{
    Dispatcher.Invoke(() => tbStatus.Text += $"GeoprocessingJob.Status={_gpJob.Status}\n");
    //if (_gpJob.Messages.Count > 0)
    //Dispatcher.Invoke(() => tbStatus.Text += $"GeoprocessingJob.Message={_gpJob.Messages[_gpJob.Messages.Count - 1].Message}\n");
    //Return if not succeeded.
    if (_gpJob.Status != JobStatus.Succeeded)
        return;
    //Get the URL to the map service.
    string gpServiceResultUrl = _lgpService.Url.ToString();
    //Get the URL segment for the specific job results.
    string jobSegment = "MapServer/jobs/" + _gpJob.ServerJobId;
    //Update the URL to point to the specific job from the service.
    gpServiceResultUrl = gpServiceResultUrl.Replace("GPServer", jobSegment);
    ArcGISMapImageLayer mil = new ArcGISMapImageLayer(new Uri(gpServiceResultUrl));
    //Load the layer.
    await mil.LoadAsync();
    // This is needed because the event comes from outside of the UI thread.
    Dispatcher.Invoke(() =>
    {
        //Add the layer to the map.
        mapView1.Map.OperationalLayers.Add(mil);
    });
}
```

重置按钮，移除地理处理结果：
```csharp
private void btReset_Click(object sender, RoutedEventArgs e)
{
    //Remove the contour.
    mapView1.Map.OperationalLayers.RemoveAt(1);
}
```

【实验结果】

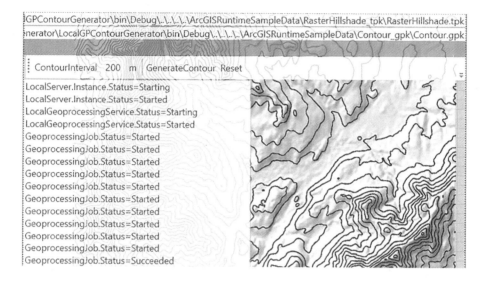

第9章 项目教学与实习案例

在学习了前面章节的小实验后,一般安排 2 周左右的集中实习,基于 ArcGIS Runtime 开发一个完整的行业地理信息系统、常用的综合工具等。在长期的项目教学和科研实践过程中,我们积累了丰富的经验,本章选取一些具有典型代表性的案例,供读者参考。

项目1 畜禽养殖污染管理地理信息系统

1. 项目简介

畜禽养殖污染管理地理信息系统是以地理信息系统为平台,不仅对养殖企业的养殖规模和污染物产生、处理、排放信息进行系统化管理,还可以对养殖企业进行定位管理。

2. 功能与创新点

对生猪、奶牛、肉牛、蛋鸡、肉鸡等养殖企业及其污染排放进行基于地理信息的空间定位管理和图表可视化展示,使其排污量及对环境的影响一目了然,便于决策者和管理人员对养殖分区划分,提出减排措施。

3. 数据源

(1)在线底图。
(2)基础地理数据。
(3)养殖行业专题数据。

4. 运行效果与项目成果

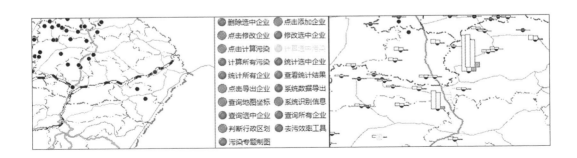

项目2 森林经营决策地理信息系统

1. 项目简介

综合化、智能化、虚拟化的林业信息集成系统，应用决策科学和森林资源经营管理的理论和方法，实现景观和林分水平上的数字化管理，为森林可持续经营提供决策支持。

2. 功能与创新点

输入：导入 GPS、RS 和 GIS 以及传感器实时监测的数据。
查看：电子林相图的浏览、导航、小班选择、图文互查。
输出：制作专题图，报表和多种矢量格式和导出栅格格式。
经营决策：森林采伐(采哪里、采多少、怎么采)、恢复、保护。
火警预报和指挥调度：通过传感器测定传入实时信息，进行预警预报，火情发生时，提供最优路线和车辆、人员调度支持。
病虫害监测与防治：病虫害预警预报，专家系统给出防治措施。
虚拟现实：对森林景观进行三维可视化显示。

3. 数据源

(1)数字高程模型。
(2)在线网络地图。
(3)林业二调数据，也可以包含一调、三调数据。

4. 运行效果和项目成果

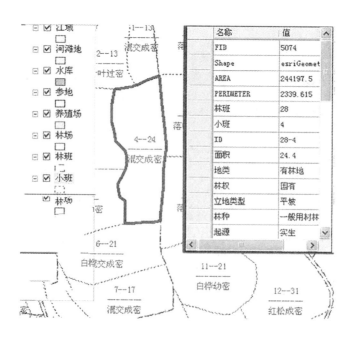

项目3 三湘四水——湖南地貌水系三维地图秀

1. 项目简介

三湘四水——湖南地貌水系三维地图建立了湖南省数字高程模型,生成了湖南水网分布图,并可叠加航空航天遥感影像,可在线或离线浏览。

2. 功能与创新点

利用水文分析在 DEM 生成水系,从宏观到微观全方位展示三湘四水丰富的地质构造、多样的地形地貌和离奇的潇湘美景,可用于科研、教学和科普宣传。

3. 数据源

(1) 数字高程模型。
(2) 基础地理信息。

4. 运行效果和项目成果

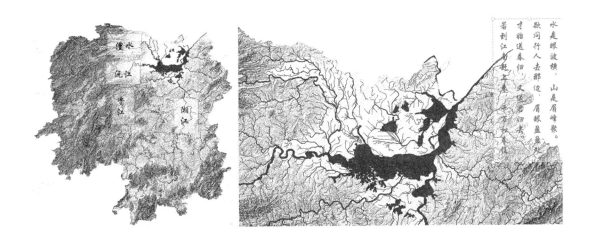

项目4 海产品溯源监测地理信息系统

1. 项目简介

通过对海产品进行扫码识别、输入查询、海产品介绍、质量监控、路径回访等功能,实现海产品养殖、生产加工、物流、售卖和消费全过程的溯源,从而保证海产品质量安全,为建立海产品质量检测体系提供技术支撑。

2. 功能与创新点

(1)扫码识别:通过扫二维码,在地图显示海产品产地,并标注产品的类别、名称等。

(2)海产品介绍:点击海产品,可获得该产品的详细介绍。

(3)质量监控:海产品养殖的环境及质量报告和产品加工后的质量报告。

(4)溯源:渔船出海捕捞轨迹、渔船到港分拣、封箱冷护、物流运输信息及产品到厂清洗、加工、包装到售卖等环节的跟踪记录。

3. 数据源

(1)在线底图。
(2)海产品专题时态数据(点类型)。

4. 运行效果和项目成果

项目 5　POI 标注地理信息系统

1. 项目简介

POI(Point Of Interest,兴趣点)主要指人们日常生活中经常遇到的地理场所,如学校、医院、宾馆、餐饮、景点和标志性建筑物等,在智能交通、应急指挥、公共安全、物流管理、电子商务以及其他各类 LBS 服务领域发挥着重要作用。

POI 是地图服务重要的矢量化形式表达方式,也是地图最鲜活的"血液",与面向公众的基于位置服务密切相关,它代表一类真实地理实体的地理空间数据。用户通过自由标注,在地图上添加兴趣点,可增强公众地图参与度,让地图焕发永久活力。

2. 功能与创新点

(1)基础信息:餐馆的名称,地理位置,联系方式。

(2)增值信息:例如餐饮菜系及口味、停车位、平均消费、推荐菜、营业时间、容纳人数、卫生状况、好评率。菜馆所属的菜系,如鲁、川、粤、闽、苏、浙、湘、徽等,不同菜系对应的口味,包括停车位信息,每位客户平均消费价位等。

(3)深度动态信息:折扣信息、最新推出的菜品和优惠信息、优惠券出处、消费者对餐馆的评论。

3. 数据源

基础底图。

4. 运行效果与项目成果

项目 6　停车场智能管理地理信息系统

1. 项目简介

基于地理信息系统建立智能化停车场管理地理信息系统,维护车辆进出的秩序,保障车辆存放的安全性、车辆存放管理的有偿性,在现代停车场管理中发挥着越来越重要的作用。

2. 功能与创新点

(1) 停车场信息管理：对停车场位置、价格、停车位分布、停车位状态进行发布。
(2) 车辆管理：车辆停放位置、时长等。

3. 数据源

(1) 在线底图。
(2) 道路和导航信息。
(3) 停车场专题信息。

4. 运行效果与项目成果

停车场地址	向日葵收费停车场
停车场电话	19977930100
车位编号	A006
车牌号码	湘 A-12345
进入时间	2018/11/11
停车时长	2：10
停车费用	10 元

项目 7　不动产登记地理信息系统

1. 项目简介

不动产登记是《中华人民共和国物权法》确立的一项物权制度，是指经权利人或利害关系人申请，由国家专职部门将有关不动产物权及其变动事项记载于不动产登记簿的事实。将土地、房屋、林地、农地进行统一管理、统一登记。改变土地、房屋、林地和农地等分开管理的模式，按空间位置关系和统一标准，应用二维、三维 GIS 技术，统一整合土地、林地、草原等各类信息，形成集体土地所有权、房屋构筑物所有权等统一的以土地登记为核心的不动产产权"一张图",基于"一张图"实现不动产"以图管理"、"立体化管理"。

2. 功能与创新点

不动产登记申请、不动产登记簿录入、不动产籍图。

3. 数据源

(1)基础地理信息。
(2)数字高程模型。
(3)不动产专题数据。

4. 运行效果与界面成果

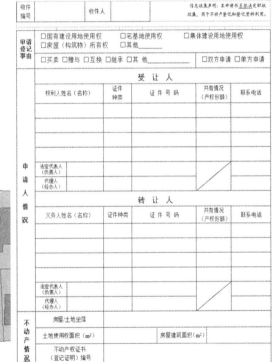

项目 8 通用地理信息平台

1. 项目简介

开发一个通用的地理信息平台。

2. 功能与创新点

(1) 绿色、免安装、轻量级。
(2) 多源数据支持,包括本地 SHP、gcodatabase、TIF 等。
(3) 支持云计算,可加载在线网络地图。
(4) 通过迭代开发,逐步丰富支持的数据格式,完善工具箱的功能。

3. 界面设计

参考 ArcMap,包括菜单、工具栏、图层列表、地图显示区、工具箱和状态栏等。

项目9　通用三维地理信息系统

1. 项目简介

开发一个通用的三维信息平台。

2. 功能与创新点

(1) 二维、三维联动。
(2) 街景播放(音频、视频、语音、文字)的功能。

3. 界面设计

参考 ArcGIS Pro。

项目10　万能地图下载器

1. 项目简介

通过网络地图服务,开发通用的网络地图浏览和下载器。

2. 功能与创新点

(1) 多源在线地图。
(2) 本地数据支持。

3. 数据源

(1) 在线底图,如 ESRI Online、百度地图、高德地图、Bing 地图、Google 地图。
(2) 本地 SHP、KML(KMZ),用于确定感兴趣区。

4. 界面设计

参考水经注、91卫图、Localspaceviewer 等网络地图平台。

项目 11 基于深度学习的智能解译系统

1. 项目简介

用户点击或者绘制多边形选择感兴趣区域，利用深度学习识别出处目标。

2. 功能与创新点

(1) 即点即识。
(2) 集成深度学习。

3. 数据源

(1) 本地 TIF 影像。
(2) 在线公众互联网地图。

项目 12 移动导航

1. 项目简介

智能移动终端上的导航 App。

2. 功能与创新点

(1) 实时更新设备的地理位置。
(2) 最短路径候选和导航。
(3) 方便的兴趣点、轨迹导入导出。

3. 数据源

在线底图服务。
本地 SHP、KML 支持。

4. 界面设计

参考高德地图 App、百度地图 App，同时加入本地数据支持。

项目 13 遥感解译外业数据采集

1. 项目简介

基于大比例尺(如 1∶500)高分辨率遥感影像进行野外实地调查和内业建库是现代空间数据采集的主要方式。例如土地确权、草地资源调查、林业资源调查、国土资源二调三调等，均大量采用此方式。开发外业数据采集和内业建库 App，可从 Portal 服务器获取数据与服务。

2. 功能与创新点

(1) 利用模版创建点、线、面，图形属性编辑，支持现场勾绘。
(2) 具有测量、数据离线编辑与同步、图层管理、定位定向等基本功能。
(3) 客户端相互通讯、数据共享、位置共享。

3. 数据源

(1) 影像切片，用于调查底图。
(2) 本地矢量数据，用于编辑。

项目 14　出租车监控服务端和客户端

1. 项目简介

对城市出租车进行轨迹记录、监控、调度和预警。

2. 功能与创新点

(1) 服务端：出租车动态监控。
(2) 服务端：地理围栏预警。
(3) 客户端：路径分析与导航。
(4) 客户端与服务端的双向通信；以及客户端与客户端的相互沟通，位置共享。

3. 数据源

(1) 在线底图服务。
(2) 车辆实时地理位置更新。

项目 15　导游导览导流预警智慧景区

1. 项目简介

收集实时游客数量、天气等数据，以地图、音频、视频为载体，以游客承载力模型为依据，针对游客提供个性化的语音导游、路线导览、空闲景点推荐以及超限游客预警等智慧服务。

2. 功能与创新点

(1) 导游：旅游景点分区和介绍。
(2) 导航：景点路线导航。
(3) 导览：空闲景点推荐。
(4) 预警：景点人口承载力、超员预警，快速出口推荐。

(5)轨迹：基于大数据展示游客轨迹。

3. 数据源

(1)景区导览底图。
(2)景区范围。
(3)景点介绍(视频、音频、文字)。
(4)景点承载力。
(5)物联网数据接口。

项目16　公共设施管理系统

1. 项目简介

对路灯、井盖等公共设施进行定位管理，可以提高管理效率和公众参与感。

2. 功能与创新点

(1)公共设施的定位展示。
(2)报修单填写。
(3)现场拍照举证。

3. 数据源

公共设施的分布点位和属性。

参 考 文 献

ESRI. ArcGIS Runtime Developer Guide（ArcGIS Runtime API for .NET）. https://developers.arcgis.com/net/